Mometrix
TEST PREPARATION

Common Core High School: Statistics and Probability Secrets Study Guide

DEAR FUTURE EXAM SUCCESS STORY

First of all, **THANK YOU** for purchasing Mometrix study materials!

Second, congratulations! You are one of the few determined test-takers who are committed to doing whatever it takes to excel on your exam. **You have come to the right place.** We developed these study materials with one goal in mind: to deliver you the information you need in a format that's concise and easy to use.

In addition to optimizing your guide for the content of the test, we've outlined our recommended steps for breaking down the preparation process into small, attainable goals so you can make sure you stay on track.

We've also analyzed the entire test-taking process, identifying the most common pitfalls and showing how you can overcome them and be ready for any curveball the test throws you.

Standardized testing is one of the biggest obstacles on your road to success, which only increases the importance of doing well in the high-pressure, high-stakes environment of test day. Your results on this test could have a significant impact on your future, and this guide provides the information and practical advice to help you achieve your full potential on test day.

Your success is our success

We would love to hear from you! If you would like to share the story of your exam success or if you have any questions or comments in regard to our products, please contact us at **800-673-8175** or **support@mometrix.com**.

Thanks again for your business and we wish you continued success!

Sincerely,
The Mometrix Test Preparation Team

Written and edited by the Mometrix Exam Secrets Test Prep Team
Printed in the United States of America

TABLE OF CONTENTS

Introduction

Thank you for purchasing this resource! You have made the choice to prepare yourself for a test that could have a huge impact on your future, and this guide is designed to help you be fully ready for test day. Obviously, it's important to have a solid understanding of the test material, but you also need to be prepared for the unique environment and stressors of the test, so that you can perform to the best of your abilities.

For this purpose, the first section that appears in this guide is the **Secret Keys**. We've devoted countless hours to meticulously researching what works and what doesn't, and we've boiled down our findings to the five most impactful steps you can take to improve your performance on the test. We start at the beginning with study planning and move through the preparation process, all the way to the testing strategies that will help you get the most out of what you know when you're finally sitting in front of the test.

We recommend that you start preparing for your test as far in advance as possible. However, if you've bought this guide as a last-minute study resource and only have a few days before your test, we recommend that you skip over the first two Secret Keys since they address a long-term study plan.

If you struggle with **test anxiety**, we strongly encourage you to check out our recommendations for how you can overcome it. Test anxiety is a formidable foe, but it can be beaten, and we want to make sure you have the tools you need to defeat it.

Secret Key #1 – Plan Big, Study Small

There's a lot riding on your performance. If you want to ace this test, you're going to need to keep your skills sharp and the material fresh in your mind. You need a plan that lets you review everything you need to know while still fitting in your schedule. We'll break this strategy down into three categories.

Information Organization

Start with the information you already have: the official test outline. From this, you can make a complete list of all the concepts you need to cover before the test. Organize these concepts into groups that can be studied together, and create a list of any related vocabulary you need to learn so you can brush up on any difficult terms. You'll want to keep this vocabulary list handy once you actually start studying since you may need to add to it along the way.

Time Management

Once you have your set of study concepts, decide how to spread them out over the time you have left before the test. Break your study plan into small, clear goals so you have a manageable task for each day and know exactly what you're doing. Then just focus on one small step at a time. When you manage your time this way, you don't need to spend hours at a time studying. Studying a small block of content for a short period each day helps you retain information better and avoid stressing over how much you have left to do. You can relax knowing that you have a plan to cover everything in time. In order for this strategy to be effective though, you have to start studying early and stick to your schedule. Avoid the exhaustion and futility that comes from last-minute cramming!

Study Environment

The environment you study in has a big impact on your learning. Studying in a coffee shop, while probably more enjoyable, is not likely to be as fruitful as studying in a quiet room. It's important to keep distractions to a minimum. You're only planning to study for a short block of time, so make the most of it. Don't pause to check your phone or get up to find a snack. It's also important to **avoid multitasking**. Research has consistently shown that multitasking will make your studying dramatically less effective. Your study area should also be comfortable and well-lit so you don't have the distraction of straining your eyes or sitting on an uncomfortable chair.

The time of day you study is also important. You want to be rested and alert. Don't wait until just before bedtime. Study when you'll be most likely to comprehend and remember. Even better, if you know what time of day your test will be, set that time aside for study. That way your brain will be used to working on that subject at that specific time and you'll have a better chance of recalling information.

Finally, it can be helpful to team up with others who are studying for the same test. Your actual studying should be done in as isolated an environment as possible, but the work of organizing the information and setting up the study plan can be divided up. In between study sessions, you can discuss with your teammates the concepts that you're all studying and quiz each other on the details. Just be sure that your teammates are as serious about the test as you are. If you find that your study time is being replaced with social time, you might need to find a new team.

Secret Key #2 – Make Your Studying Count

You're devoting a lot of time and effort to preparing for this test, so you want to be absolutely certain it will pay off. This means doing more than just reading the content and hoping you can remember it on test day. It's important to make every minute of study count. There are two main areas you can focus on to make your studying count.

Retention

It doesn't matter how much time you study if you can't remember the material. You need to make sure you are retaining the concepts. To check your retention of the information you're learning, try recalling it at later times with minimal prompting. Try carrying around flashcards and glance at one or two from time to time or ask a friend who's also studying for the test to quiz you.

To enhance your retention, look for ways to put the information into practice so that you can apply it rather than simply recalling it. If you're using the information in practical ways, it will be much easier to remember. Similarly, it helps to solidify a concept in your mind if you're not only reading it to yourself but also explaining it to someone else. Ask a friend to let you teach them about a concept you're a little shaky on (or speak aloud to an imaginary audience if necessary). As you try to summarize, define, give examples, and answer your friend's questions, you'll understand the concepts better and they will stay with you longer. Finally, step back for a big picture view and ask yourself how each piece of information fits with the whole subject. When you link the different concepts together and see them working together as a whole, it's easier to remember the individual components.

Finally, practice showing your work on any multi-step problems, even if you're just studying. Writing out each step you take to solve a problem will help solidify the process in your mind, and you'll be more likely to remember it during the test.

Modality

Modality simply refers to the means or method by which you study. Choosing a study modality that fits your own individual learning style is crucial. No two people learn best in exactly the same way, so it's important to know your strengths and use them to your advantage.

For example, if you learn best by visualization, focus on visualizing a concept in your mind and draw an image or a diagram. Try color-coding your notes, illustrating them, or creating symbols that will trigger your mind to recall a learned concept. If you learn best by hearing or discussing information, find a study partner who learns the same way or read aloud to yourself. Think about how to put the information in your own words. Imagine that you are giving a lecture on the topic and record yourself so you can listen to it later.

For any learning style, flashcards can be helpful. Organize the information so you can take advantage of spare moments to review. Underline key words or phrases. Use different colors for different categories. Mnemonic devices (such as creating a short list in which every item starts with the same letter) can also help with retention. Find what works best for you and use it to store the information in your mind most effectively and easily.

Secret Key #3 – Practice the Right Way

Your success on test day depends not only on how many hours you put into preparing, but also on whether you prepared the right way. It's good to check along the way to see if your studying is paying off. One of the most effective ways to do this is by taking practice tests to evaluate your progress. Practice tests are useful because they show exactly where you need to improve. Every time you take a practice test, pay special attention to these three groups of questions:

- The questions you got wrong
- The questions you had to guess on, even if you guessed right
- The questions you found difficult or slow to work through

This will show you exactly what your weak areas are, and where you need to devote more study time. Ask yourself why each of these questions gave you trouble. Was it because you didn't understand the material? Was it because you didn't remember the vocabulary? Do you need more repetitions on this type of question to build speed and confidence? Dig into those questions and figure out how you can strengthen your weak areas as you go back to review the material.

Additionally, many practice tests have a section explaining the answer choices. It can be tempting to read the explanation and think that you now have a good understanding of the concept. However, an explanation likely only covers part of the question's broader context. Even if the explanation makes perfect sense, **go back and investigate** every concept related to the question until you're positive you have a thorough understanding.

As you go along, keep in mind that the practice test is just that: practice. Memorizing these questions and answers will not be very helpful on the actual test because it is unlikely to have any of the same exact questions. If you only know the right answers to the sample questions, you won't be prepared for the real thing. **Study the concepts** until you understand them fully, and then you'll be able to answer any question that shows up on the test.

It's important to wait on the practice tests until you're ready. If you take a test on your first day of study, you may be overwhelmed by the amount of material covered and how much you need to learn. Work up to it gradually.

On test day, you'll need to be prepared for answering questions, managing your time, and using the test-taking strategies you've learned. It's a lot to balance, like a mental marathon that will have a big impact on your future. Like training for a marathon, you'll need to start slowly and work your way up. When test day arrives, you'll be ready.

Start with the strategies you've read in the first two Secret Keys—plan your course and study in the way that works best for you. If you have time, consider using multiple study resources to get different approaches to the same concepts. It can be helpful to see difficult concepts from more than one angle. Then find a good source for practice tests. Many times, the test website will suggest potential study resources or provide sample tests.

Practice Test Strategy

If you're able to find at least three practice tests, we recommend this strategy:

Untimed and Open-Book Practice

Take the first test with no time constraints and with your notes and study guide handy. Take your time and focus on applying the strategies you've learned.

Timed and Open-Book Practice

Take the second practice test open-book as well, but set a timer and practice pacing yourself to finish in time.

Timed and Closed-Book Practice

Take any other practice tests as if it were test day. Set a timer and put away your study materials. Sit at a table or desk in a quiet room, imagine yourself at the testing center, and answer questions as quickly and accurately as possible.

Keep repeating timed and closed-book tests on a regular basis until you run out of practice tests or it's time for the actual test. Your mind will be ready for the schedule and stress of test day, and you'll be able to focus on recalling the material you've learned.

Secret Key #4 – Pace Yourself

Once you're fully prepared for the material on the test, your biggest challenge on test day will be managing your time. Just knowing that the clock is ticking can make you panic even if you have plenty of time left. Work on pacing yourself so you can build confidence against the time constraints of the exam. Pacing is a difficult skill to master, especially in a high-pressure environment, so **practice is vital**.

Set time expectations for your pace based on how much time is available. For example, if a section has 60 questions and the time limit is 30 minutes, you know you have to average 30 seconds or less per question in order to answer them all. Although 30 seconds is the hard limit, set 25 seconds per question as your goal, so you reserve extra time to spend on harder questions. When you budget extra time for the harder questions, you no longer have any reason to stress when those questions take longer to answer.

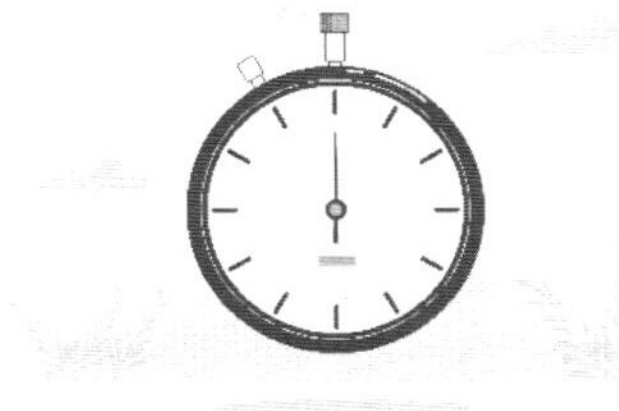

Don't let this time expectation distract you from working through the test at a calm, steady pace, but keep it in mind so you don't spend too much time on any one question. Recognize that taking extra time on one question you don't understand may keep you from answering two that you do understand later in the test. If your time limit for a question is up and you're still not sure of the answer, mark it and move on, and come back to it later if the time and the test format allow. If the testing format doesn't allow you to return to earlier questions, just make an educated guess; then put it out of your mind and move on.

On the easier questions, be careful not to rush. It may seem wise to hurry through them so you have more time for the challenging ones, but it's not worth missing one if you know the concept and just didn't take the time to read the question fully. Work efficiently but make sure you understand the question and have looked at all of the answer choices, since more than one may seem right at first.

Even if you're paying attention to the time, you may find yourself a little behind at some point. You should speed up to get back on track, but do so wisely. Don't panic; just take a few seconds less on each question until you're caught up. Don't guess without thinking, but do look through the answer choices and eliminate any you know are wrong. If you can get down to two choices, it is often worthwhile to guess from those. Once you've chosen an answer, move on and don't dwell on any that you skipped or had to hurry through. If a question was taking too long, chances are it was one of the harder ones, so you weren't as likely to get it right anyway.

On the other hand, if you find yourself getting ahead of schedule, it may be beneficial to slow down a little. The more quickly you work, the more likely you are to make a careless mistake that will affect your score. You've budgeted time for each question, so don't be afraid to spend that time. Practice an efficient but careful pace to get the most out of the time you have.

Secret Key #5 – Have a Plan for Guessing

When you're taking the test, you may find yourself stuck on a question. Some of the answer choices seem better than others, but you don't see the one answer choice that is obviously correct. What do you do?

The scenario described above is very common, yet most test takers have not effectively prepared for it. Developing and practicing a plan for guessing may be one of the single most effective uses of your time as you get ready for the exam.

In developing your plan for guessing, there are three questions to address:

- When should you start the guessing process?
- How should you narrow down the choices?
- Which answer should you choose?

When to Start the Guessing Process

Unless your plan for guessing is to select C every time (which, despite its merits, is not what we recommend), you need to leave yourself enough time to apply your answer elimination strategies. Since you have a limited amount of time for each question, that means that if you're going to give yourself the best shot at guessing correctly, you have to decide quickly whether or not you will guess.

Of course, the best-case scenario is that you don't have to guess at all, so first, see if you can answer the question based on your knowledge of the subject and basic reasoning skills. Focus on the key words in the question and try to jog your memory of related topics. Give yourself a chance to bring the knowledge to mind, but once you realize that you don't have (or you can't access) the knowledge you need to answer the question, it's time to start the guessing process.

It's almost always better to start the guessing process too early than too late. It only takes a few seconds to remember something and answer the question from knowledge. Carefully eliminating wrong answer choices takes longer. Plus, going through the process of eliminating answer choices can actually help jog your memory.

Summary: Start the guessing process as soon as you decide that you can't answer the question based on your knowledge.

How to Narrow Down the Choices

The next chapter in this book (**Test-Taking Strategies**) includes a wide range of strategies for how to approach questions and how to look for answer choices to eliminate. You will definitely want to read those carefully, practice them, and figure out which ones work best for you. Here though, we're going to address a mindset rather than a particular strategy.

Your odds of guessing an answer correctly depend on how many options you are choosing from.

Number of options left	5	4	3	2	1
Odds of guessing correctly	20%	25%	33%	50%	100%

You can see from this chart just how valuable it is to be able to eliminate incorrect answers and make an educated guess, but there are two things that many test takers do that cause them to miss out on the benefits of guessing:

- Accidentally eliminating the correct answer
- Selecting an answer based on an impression

We'll look at the first one here, and the second one in the next section.

To avoid accidentally eliminating the correct answer, we recommend a thought exercise called **the $5 challenge**. In this challenge, you only eliminate an answer choice from contention if you are willing to bet $5 on it being wrong. Why $5? Five dollars is a small but not insignificant amount of money. It's an amount you could afford to lose but wouldn't want to throw away. And while losing $5 once might not hurt too much, doing it twenty times will set you back $100. In the same way, each small decision you make—eliminating a choice here, guessing on a question there—won't by itself impact your score very much, but when you put them all together, they can make a big difference. By holding each answer choice elimination decision to a higher standard, you can reduce the risk of accidentally eliminating the correct answer.

The $5 challenge can also be applied in a positive sense: If you are willing to bet $5 that an answer choice *is* correct, go ahead and mark it as correct.

Summary: Only eliminate an answer choice if you are willing to bet $5 that it is wrong.

Which Answer to Choose

You're taking the test. You've run into a hard question and decided you'll have to guess. You've eliminated all the answer choices you're willing to bet $5 on. Now you have to pick an answer. Why do we even need to talk about this? Why can't you just pick whichever one you feel like when the time comes?

The answer to these questions is that if you don't come into the test with a plan, you'll rely on your impression to select an answer choice, and if you do that, you risk falling into a trap. The test writers know that everyone who takes their test will be guessing on some of the questions, so they intentionally write wrong answer choices to seem plausible. You still have to pick an answer though, and if the wrong answer choices are designed to look right, how can you ever be sure that you're not falling for their trap? The best solution we've found to this dilemma is to take the decision out of your hands entirely. Here is the process we recommend:

Once you've eliminated any choices that you are confident (willing to bet $5) are wrong, select the first remaining choice as your answer.

Whether you choose to select the first remaining choice, the second, or the last, the important thing is that you use some preselected standard. Using this approach guarantees that you will not be enticed into selecting an answer choice that looks right, because you are not basing your decision on how the answer choices look.

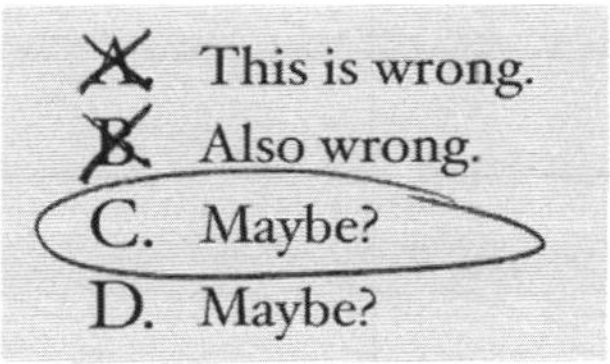

This is not meant to make you question your knowledge. Instead, it is to help you recognize the difference between your knowledge and your impressions. There's a huge difference between thinking an answer is right because of what you know, and thinking an answer is right because it looks or sounds like it should be right.

Summary: To ensure that your selection is appropriately random, make a predetermined selection from among all answer choices you have not eliminated.

Test-Taking Strategies

This section contains a list of test-taking strategies that you may find helpful as you work through the test. By taking what you know and applying logical thought, you can maximize your chances of answering any question correctly!

It is very important to realize that every question is different and every person is different: no single strategy will work on every question, and no single strategy will work for every person. That's why we've included all of them here, so you can try them out and determine which ones work best for different types of questions and which ones work best for you.

Question Strategies

✓ Read Carefully

Read the question and the answer choices carefully. Don't miss the question because you misread the terms. You have plenty of time to read each question thoroughly and make sure you understand what is being asked. Yet a happy medium must be attained, so don't waste too much time. You must read carefully and efficiently.

✓ Contextual Clues

Look for contextual clues. If the question includes a word you are not familiar with, look at the immediate context for some indication of what the word might mean. Contextual clues can often give you all the information you need to decipher the meaning of an unfamiliar word. Even if you can't determine the meaning, you may be able to narrow down the possibilities enough to make a solid guess at the answer to the question.

✓ Prefixes

If you're having trouble with a word in the question or answer choices, try dissecting it. Take advantage of every clue that the word might include. Prefixes can be a huge help. Usually, they allow you to determine a basic meaning. *Pre-* means before, *post-* means after, *pro-* is positive, *de-* is negative. From prefixes, you can get an idea of the general meaning of the word and try to put it into context.

✓ Hedge Words

Watch out for critical hedge words, such as *likely, may, can, often, almost, mostly, usually, generally, rarely*, and *sometimes*. Question writers insert these hedge phrases to cover every possibility. Often an answer choice will be wrong simply because it leaves no room for exception. Be on guard for answer choices that have definitive words such as *exactly* and *always*.

✓ Switchback Words

Stay alert for *switchbacks*. These are the words and phrases frequently used to alert you to shifts in thought. The most common switchback words are *but, although*, and *however*. Others include *nevertheless, on the other hand, even though, while, in spite of, despite*, and *regardless of*. Switchback words are important to catch because they can change the direction of the question or an answer choice.

⊘ Face Value

When in doubt, use common sense. Accept the situation in the problem at face value. Don't read too much into it. These problems will not require you to make wild assumptions. If you have to go beyond creativity and warp time or space in order to have an answer choice fit the question, then you should move on and consider the other answer choices. These are normal problems rooted in reality. The applicable relationship or explanation may not be readily apparent, but it is there for you to figure out. Use your common sense to interpret anything that isn't clear.

Answer Choice Strategies

⊘ Answer Selection

The most thorough way to pick an answer choice is to identify and eliminate wrong answers until only one is left, then confirm it is the correct answer. Sometimes an answer choice may immediately seem right, but be careful. The test writers will usually put more than one reasonable answer choice on each question, so take a second to read all of them and make sure that the other choices are not equally obvious. As long as you have time left, it is better to read every answer choice than to pick the first one that looks right without checking the others.

⊘ Answer Choice Families

An answer choice family consists of two (in rare cases, three) answer choices that are very similar in construction and cannot all be true at the same time. If you see two answer choices that are direct opposites or parallels, one of them is usually the correct answer. For instance, if one answer choice says that quantity *x* increases and another either says that quantity *x* decreases (opposite) or says that quantity *y* increases (parallel), then those answer choices would fall into the same family. An answer choice that doesn't match the construction of the answer choice family is more likely to be incorrect. Most questions will not have answer choice families, but when they do appear, you should be prepared to recognize them.

⊘ Eliminate Answers

Eliminate answer choices as soon as you realize they are wrong, but make sure you consider all possibilities. If you are eliminating answer choices and realize that the last one you are left with is also wrong, don't panic. Start over and consider each choice again. There may be something you missed the first time that you will realize on the second pass.

⊘ Avoid Fact Traps

Don't be distracted by an answer choice that is factually true but doesn't answer the question. You are looking for the choice that answers the question. Stay focused on what the question is asking for so you don't accidentally pick an answer that is true but incorrect. Always go back to the question and make sure the answer choice you've selected actually answers the question and is not merely a true statement.

⊘ Extreme Statements

In general, you should avoid answers that put forth extreme actions as standard practice or proclaim controversial ideas as established fact. An answer choice that states the "process should be used in certain situations, if..." is much more likely to be correct than one that states the "process should be discontinued completely." The first is a calm rational statement and doesn't even make a definitive, uncompromising stance, using a hedge word *if* to provide wiggle room, whereas the second choice is far more extreme.

⊘ Benchmark

As you read through the answer choices and you come across one that seems to answer the question well, mentally select that answer choice. This is not your final answer, but it's the one that will help you evaluate the other answer choices. The one that you selected is your benchmark or standard for judging each of the other answer choices. Every other answer choice must be compared to your benchmark. That choice is correct until proven otherwise by another answer choice beating it. If you find a better answer, then that one becomes your new benchmark. Once you've decided that no other choice answers the question as well as your benchmark, you have your final answer.

⊘ Predict the Answer

Before you even start looking at the answer choices, it is often best to try to predict the answer. When you come up with the answer on your own, it is easier to avoid distractions and traps because you will know exactly what to look for. The right answer choice is unlikely to be word-for-word what you came up with, but it should be a close match. Even if you are confident that you have the right answer, you should still take the time to read each option before moving on.

General Strategies

⊘ Tough Questions

If you are stumped on a problem or it appears too hard or too difficult, don't waste time. Move on! Remember though, if you can quickly check for obviously incorrect answer choices, your chances of guessing correctly are greatly improved. Before you completely give up, at least try to knock out a couple of possible answers. Eliminate what you can and then guess at the remaining answer choices before moving on.

⊘ Check Your Work

Since you will probably not know every term listed and the answer to every question, it is important that you get credit for the ones that you do know. Don't miss any questions through careless mistakes. If at all possible, try to take a second to look back over your answer selection and make sure you've selected the correct answer choice and haven't made a costly careless mistake (such as marking an answer choice that you didn't mean to mark). This quick double check should more than pay for itself in caught mistakes for the time it costs.

⊘ Pace Yourself

It's easy to be overwhelmed when you're looking at a page full of questions; your mind is confused and full of random thoughts, and the clock is ticking down faster than you would like. Calm down and maintain the pace that you have set for yourself. Especially as you get down to the last few minutes of the test, don't let the small numbers on the clock make you panic. As long as you are on track by monitoring your pace, you are guaranteed to have time for each question.

⊘ Don't Rush

It is very easy to make errors when you are in a hurry. Maintaining a fast pace in answering questions is pointless if it makes you miss questions that you would have gotten right otherwise. Test writers like to include distracting information and wrong answers that seem right. Taking a little extra time to avoid careless mistakes can make all the difference in your test score. Find a pace that allows you to be confident in the answers that you select.

⊘ KEEP MOVING

Panicking will not help you pass the test, so do your best to stay calm and keep moving. Taking deep breaths and going through the answer elimination steps you practiced can help to break through a stress barrier and keep your pace.

Final Notes

The combination of a solid foundation of content knowledge and the confidence that comes from practicing your plan for applying that knowledge is the key to maximizing your performance on test day. As your foundation of content knowledge is built up and strengthened, you'll find that the strategies included in this chapter become more and more effective in helping you quickly sift through the distractions and traps of the test to isolate the correct answer.

Now that you're preparing to move forward into the test content chapters of this book, be sure to keep your goal in mind. As you read, think about how you will be able to apply this information on the test. If you've already seen sample questions for the test and you have an idea of the question format and style, try to come up with questions of your own that you can answer based on what you're reading. This will give you valuable practice applying your knowledge in the same ways you can expect to on test day.

Good luck and good studying!

Statistics and Probability

DOT PLOT, HISTOGRAM, AND BOX PLOT

Dot plots, histograms, and box plots are three different methods of representing data sets in graphical form. While they all have similar purposes, some are more suitable for some data sets than others. Dot plots represent each data point as a separate dot, and are most useful for relatively small data sets with a small number of possible values. When the data set contains more than a few dozen points, dot plots may become unwieldy. For larger data sets, where the data distribution is continuous rather than discrete, histograms may be more practical. Histograms are especially useful to estimate the density of the data, and to estimate probability density functions. Box plots may be used to summarize the shape of a larger number of univariate data values; they are most useful for comparing separate groups of data, such as the results of several different experiments, or the statistics of several discrete populations.

DOT PLOT

A dot plot is a representation of a data set in which each data point is represented by a dot or similar marking, with matching data points grouped in columns. For instance, the data set {2, 1, 3, 1, 1, 5, 4, 4, 3, 3, 3, 4, 1, 5} can be represented as a dot plot as follows:

For large data sets, it is possible for each dot in a dot plot to represent more than one data point. However, for such data sets other representations may be more suitable.

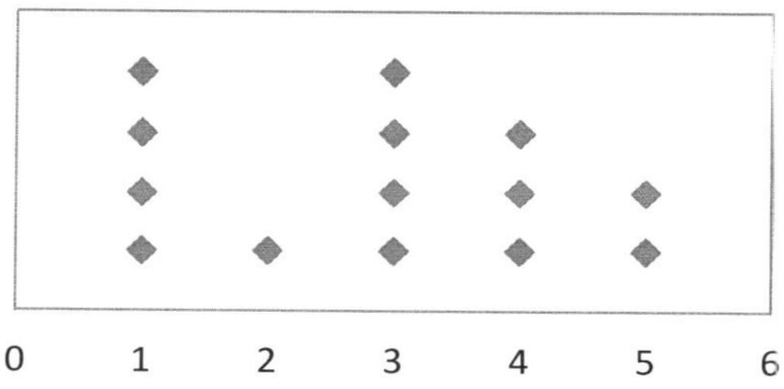

HISTOGRAM

A histogram is a representation of a data set in which the data are represented by bars corresponding to discrete intervals, with the height of each bar representative of the number of data points falling in the corresponding interval. For example, the data set {101, 141, 105, 159, 122, 107, 145, 153, 183, 172, 164, 162, 144, 132, 138, 116, 155, 147, 141, 129, 168, 145, 152} can be represented by a histogram as follows:

Histograms are useful for large data sets with values that may range continuously over intervals rather than being confined to discrete possibilities.

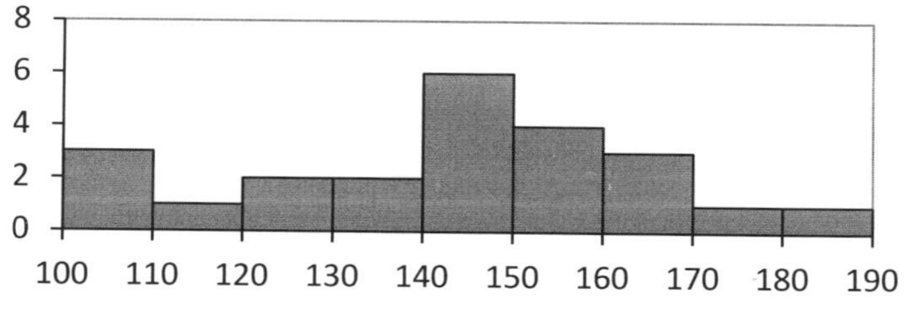

Box Plot

A *box plot*, or box-and-whisker plot, is a representation of one or more data sets in which each data set is represented by a box with a bar in the middle and a "whisker" on each side. The bar represents the *median* of the data, and the edges of the box represent the first and third quartiles. The ends of the whiskers may represent the maximum and minimum data values, although often outliers are excluded and are represented instead as discrete points.

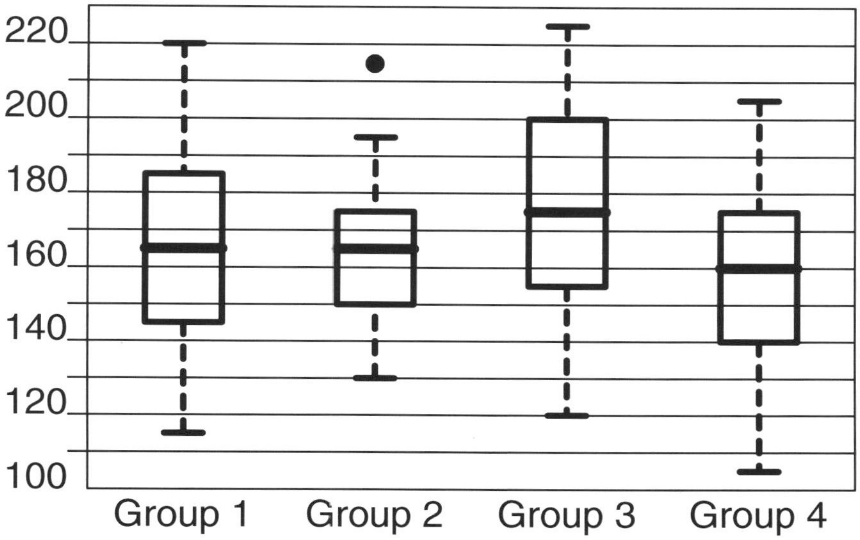

For example, given the data, 5, 5, 6, 9, 12, 14, 15, 17, 17, 21, 24, 26, 29, 31, 36, 38, 39, 46, 47, 49, the following summary statistics may be recorded: Median = 22.5, Q1 = 13, Q3 = 37, minimum = 5, and maximum = 49. Thus, a box plot of this data will show a box, with a middle bar at the value, 22.5, edges of the box at the values, 13 and 37, and whiskers at the values, 5 and 49. Box plots are useful when it is desired to compare the statistics of multiple related data sets, such as several different groups of experimental subjects.

Measures of Central Tendency

Mean

The mean is the average of the data points; that is, it is the sum of the data points divided by the number of data points. Mathematically, the mean of a set of data points $\{x_1, x_2, x_3, \ldots x_n\}$can be written as$\bar{X} = \sum \frac{X}{N}$. For instance, for the data set (1, 3, 6, 8, 100, 800), the mean is $\frac{1+3+6+8+100+800}{6} = 153$.

The mean is most useful, when data is approximately normal and does not include extreme outliers. In the above example, the data shows much variation. Thus, the mean is not the best measure of central tendency to use, when interpreting the data. With this data set, the median will give a more complete picture of the distribution.

Median

The median is the value in the middle of the data set, in the sense that 50% of the data points lie above the median and 50% of the data points lie below. The median can be determined by simply putting the data points in order, and selecting the data point in the middle. If there is an even number of data points, then the median is the average of the middle two data points. For instance, for the data set {1, 3, 6, 8, 100, 800}, the median is $\frac{6+8}{2} = 7$.

For distributions with widely varying data points, especially those with large outliers, the median is a more appropriate measure of central tendency, and thus gives a better idea of a "typical" data point. Notice in the data set above, the mean is 153, while the median is 7.

MODE

The mode is the value that appears most often in the data set. For instance, for the data set {2, 6, 4, 9, 4, 5, 7, 6, 4, 1, 5, 6, 7, 5, 6}, the mode is 6: the number 6 appears four times in the data set, while the next most frequent values, 4 and 5, appear only three times each. It is possible for a data set to have more than one mode: in the data set {11, 14, 17, 16, 11, 17, 12, 14, 17, 14, 13}, 14 and 17 are both modes, appearing three times each. In the extreme case of a uniform distribution—a distribution in which all values appear with equal probability—all values in the data set are modes.

The mode is useful to get a general sense of the shape of the distribution; it shows where the peaks of the distribution are. More information is necessary to get a more detailed description of the full shape.

RELATION TO SHAPE OF MEASUREMENT

The measurement of central tendency with the most clearly visible relationship to the shape is the mode. The mode defines the peak of the distribution, and a distribution with multiple modes has multiple peaks. The relationship of the shape to the other measurements of central tendency is more subtle. For a symmetrical distribution with a single peak, the mode, median, and mean all coincide. For a distribution skewed to the left or right, however, this is not generally the case. One rule of thumb often given is that the median is displaced from the mode in the same direction as the skew of the graph, and the mean in the same direction farther still.

FIRST QUARTILE

The first quartile of a data set is a value greater than or equal to one quarter of the data points (and less than the other three quarters). Various methods exist for defining the first quartile precisely; one of the simplest is to define the first quartile as the median of the first half of the ordered data (excluding the median if there are an odd number of data points).

Applying this method, for example, to the data set {3, 1, 12, 7, 17, 4, 10, 8, 9, 20, 4}, we proceed as follows: Putting the data in order, we get {1, 3, 4, 4, 7, 8, 9, 10, 12, 17, 20}. The first half (excluding the median) is {1, 3, 4, 4, 7}, which has a median of 4. Therefore the first quartile of this data set is 4.

THIRD QUARTILE

The third quartile of a data set is a value greater than or equal to three quarters of the data points (and less than the remaining quarter). Various methods exist for defining the third quartiles precisely; one of the simplest is to define the third quartile as the median of the second half of the ordered data (excluding the median if there are an odd number of data points). Applying this method, for example, to the data set {3, 1, 12, 7, 17, 4, 10, 8, 9, 20, 4}, we proceed as follows: Putting the data in order, we get {1, 3, 4, 4, 7, 8, 9, 10, 12, 17, 20}. The second half (excluding the median) is {9, 10, 12, 17, 20}, which has a median of 12. Therefore the third quartile of this data set is 12.

INTERQUARTILE RANGE OF A DATA SET

The interquartile range of a data set is the difference between the third and first quartiles. That is, one quarter of the data fall below the interquartile range and one quarter of the data above it. Exactly half of the data points fall within the interquartile range, half of those above the median and half below. (This is, of course, why the quartile points are called "quartiles", because they divide the data into quarters: one quarter of the data points are below the first quartile, one quarter between the first and second quartile (the median), and so on.) The interquartile range is useful to get a rough idea of the spread of the data. The median by itself shows where the data are centered (or rather, shows one measure of central tendency); the interquartile range gives a better idea of how much the data points vary from this center.

STANDARD DEVIATION

The standard deviation of a data set is a measurement of how much the data points vary from the mean. More precisely, it is equal to the square root of the average of the squares of the differences between each point and the mean:$s_x = \sqrt{\frac{\sum(X-\bar{X})^2}{N-1}}$.

The standard deviation is useful for determining the spread, or dispersion, of the data, or how far they vary from the mean. The smaller the standard deviation, the closer the values tend to be to the mean; the larger the standard deviation, the more they tend to be scattered far from the mean.

OUTLIER

An outlier is an extremely high or extremely low value in the data set. It may be the result of measurement error, in which case, the outlier is not a valid member of the data set. However, it may also be a valid member of the distribution. Unless a measurement error is identified, the experimenter cannot know for certain if an outlier is or is not a member of the distribution. There are arbitrary methods that can be employed to designate an extreme value as an outlier. One method designates an outlier (or possible outlier) to be any value less than $Q_1 - 1.5(IQR)$ or any value greater than $Q_3 + 1.5(IQR)$, where Q_1 and Q_3 are the first and third quartiles and IQR is the interquartile range. For instance, in the data set {42, 71, 22, 500, 33, 38, 62, 44, 58, 37, 61, 25}, the point 500 may be considered an outlier, since 500 is greater than 101.25 (61.5 + 1.5(26.5) = 101.25).

EFFECT ON MEASUREMENTS OF THE CENTER AND SPREAD OF A DATA DISTRIBUTION

Regarding measurements of the center, outliers tend to have little or no effect on the mode, and in general little effect on the median, though their effects may be magnified for very small distributions. Means, however, are very sensitive to outliers; a single data point that lies far outside the range of the others may leave the mode and median almost unchanged while drastically altering the mean. For instance, the data set {2, 2, 5, 5, 5, 8, 11} has a mode and median of 5 and a mean of approximately 5.4; adding the outlying value 650 to the data set leaves the mode and median unchanged but increases the mean to 86.

Like the median, the interquartile range is little affected by outliers, though, again, the effect may be greater for small data sets. The standard deviation, like the mean, is much more sensitive to outliers, and may be significantly increased by a single outlier that lies far from the spread of the rest of the points.

TWO DATA SETS WITH THE SAME CENTER BUT DIFFERENT SPREADS

It's certainly possible for two data sets to have the same center but different spreads; all that's necessary is for the points in one set to be clustered closer to the center than the other. Consider, for instance, two data sets that both have the normal distribution with the same mean but different standard deviations: all usual measurements of central tendency would match—the mean, mode, and median—but their spread, as measured by either interquartile range or by standard deviation, would differ. The graph below shows the distributions of two such data sets:

Two Data Sets with the Same Spread but Different Centers

It's certainly possible for two data sets to have the same spread but different centers; nothing prevents the data points of two sets from being equally near the center even if the center differs. Consider, for instance, two data sets that both have the normal distribution with the same standard deviation but different means: their spread, as measured by either interquartile range or by standard deviation, would match, but the measures of central tendency, mean, mode, and median—would differ between the data sets. The graph below shows the distributions of two such data sets:

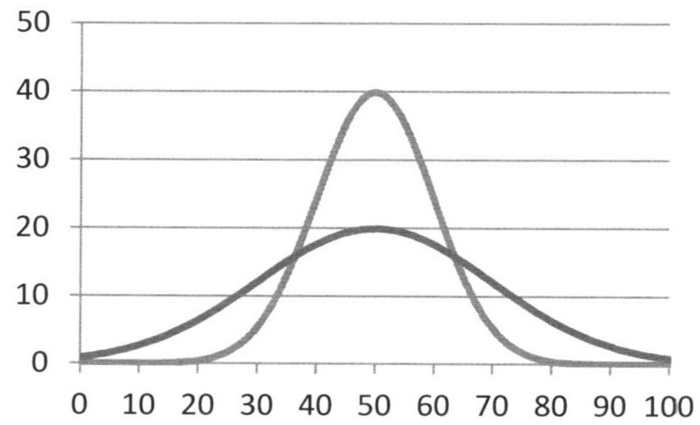

Two Data Distributions with the Same Spread and Center but Different Shapes

It's certainly possible for two data distributions to have the same spread and center but different shapes. While the spread and the center are significant characteristics of a distribution, they are not sufficient to uniquely determine the distribution; two distributions may very well have the same spread and center but differ in the details. Consider, for instance, the data sets {5, 10, 15, 20, 25, 30, 35} and {7, 10, 11, 20, 29, 30, 33}: both these data sets have a mean and median of 20, an interquartile range of 20, and a standard deviation of approximately 10.8; yet clearly they are not identical, and have different shapes. For another example, consider an asymmetric distribution reflected about its mean; despite their different shapes, the distribution and its reflection have the same mean and necessarily the same spread.

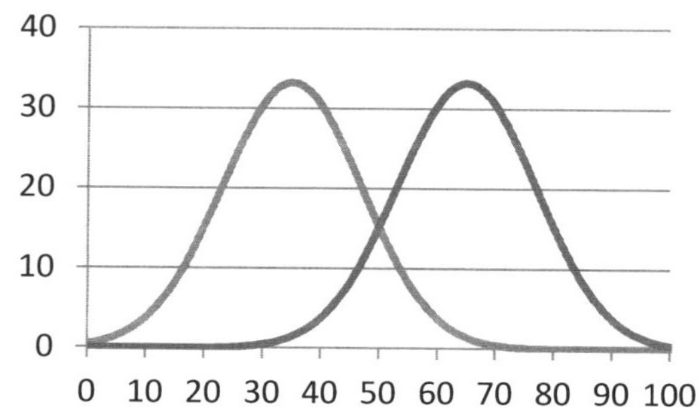

Normal Distribution

A normal distribution is a symmetrical, bell-shaped distribution that can be used to model real-world situations. In particular, the normal distribution is a good match for data that represents the sum or the mean of a large number of similar variables acting independently. If you roll a large number of dice, for instance, their mean will tend to follow close to a normal distribution. The normal distribution has a number of useful properties. Perhaps the most notable is that the distribution of the sum or difference of two variables each of which follows the normal distributions is itself another normal distribution, with the mean equal to the sum or difference of the means of the distributions of the original variables, and the standard deviation equal to the square root of the sum of the squares of the standard deviations of the original distributions.

Circumstances Not Appropriate for Normal Distribution

Many data sets that arise in real-world problems can be fit to or at least approximated by a normal distribution, and given the useful properties of this distribution it's often useful to do so. However, the normal distribution is not a good fit for all data sets. In general, if a data set seems to follow a symmetrical bell curve, it is likely (though not necessarily the case) that it can be usefully fit to a normal distribution. For a data sets that is clearly skewed and asymmetrical, however, such a fit is not suitable. Nor is it appropriate to try to fit to a normal distribution data sets that have no peaks or multiple peaks (though it may be possible to fit such data to a sum of normal distributions).

Generating Normal Distribution from the Given Mean and Standard Deviation

The normal distribution is defined by the normal equation, $f(x) = \frac{1}{\sigma\sqrt{2\pi}} e^{-\frac{1}{2}\left(\frac{x-\mu}{\sigma}\right)^2}$, where μ is the mean and σ is the standard deviation. Generating a normal equation corresponding to a given mean and standard deviation, then, is as simple as putting the appropriate values for μ and σ into this equation. A standard deviation with a mean of 100 and a standard deviation of 10, for instance, would have a normal equation $f(x) = \frac{1}{10\sqrt{2\pi}} e^{-\frac{1}{2}\left(\frac{x-100}{10}\right)^2}$. The probability that the data lies within a certain range can be found by determining the area under the normal curve—the graph of the normal function—within the given range. It's often simpler, however, to consider the standard normal distribution, with the equation $f(z) = \frac{1}{\sqrt{2\pi}} e^{-\frac{1}{2}z^2}$. This is simply a normal distribution with a mean of 0 and a standard deviation of 1. A standard z-distribution is represented by the formula, $z = \frac{X-\mu}{\sigma}$.

Estimating Population Percentages

The population percentages falling within certain ranges in a normal distribution can be estimated by finding the area under the normal curve. This can either be estimated by inspection or determined using a calculator. For certain values, however, the population percentages can be estimated more directly. About 68% of the data points in a normal distribution lie within one standard deviation of the mean, about 95% within two standard deviations, and about 99.9% within three. For a normal distribution with a mean of 100 and a standard deviation of 10, for instance, we would expect 68% of the data points to lie between 90 and 110 (100 ± 10), and 95% of the data points to lie between 80 and 120. Because of the normal distribution's symmetry, half of these would lie on each side of the mean, so, for instance, about 34% of the data points would lie between 90 and 100 and 34% between 100 and 110; about 2.5% of the data points (½(100% – 95%)) would exceed 120.

ESTIMATING AREA UNDER A NORMAL CURVE

USING A CALCULATOR

Most scientific calculators have statistical functions that allow the easy calculation of probabilities related to common probability distributions, including the normal distribution. The details depend on the particular model of calculator. In the TI-84, one of the most commonly used calculators today, the appropriate function can be accessed from the DISTR button (press "2nd", and then "VARS"). This will bring up a menu of distribution-related functions; select "normalcdf". The parameters of this function are, in order, the lower bound, the upper bound, the mean, and the standard deviation. For a normal distribution with a mean of 100 and a standard deviation of 10, for instance, to find the probability that a data point lies between 105 and 115 you would enter "normalcdf(105, 115, 100, 10)", yielding an answer of about 0.242, or 24.2%. If you desire to see the area visually, you can hit the right arrow from the DISTR menu to get to the DRAW menu, and choose "ShadeNorm". The parameters are the same as for the "normalcdf" function.

USING A SPREADSHEET

Most modern spreadsheet programs include functions that allow the user to find the area under a normal curve. In Microsoft Excel, the appropriate function is NORMDIST, which gives the total area of the graph to the left of a particular value. The first parameter of this function is the value in question, the second is the mean, and the third the standard deviation. The fourth parameter should be simply set to TRUE. (If it's set to FALSE, Excel will return the value of the normal function at that point rather than the area under the curve.) For instance, for a normal distribution with a mean of 100 and a standard deviation of 10, to find the area under the curve to the left of 115, you would enter into the cell "=NORMDIST(115,100,10,TRUE)".

To find the area within a given interval, you can simply take the difference of two results. For instance, for the previous example, the area of the curve between 105 and 115 would be generated by "=NORMDIST(115,100,10,TRUE)-NORMDIST(105,100,10,TRUE)".

USING A TABLE

Z-tables are available and give the total area under a normal curve to the left of a given value (or the larger portion). The area within an interval can be found by taking the difference between the areas for the right and left endpoints. A mean to z table is also available and may be used to find the area between the mean and a given value. This table may also be used, in order to find the area to the left (or right) or a value, by adding half the area under the normal curve (0.5), to the mean to z area represented by the given value.

Z-tables represent standardized values. The raw values are not represented by the table. Instead, the table represents standardized z-scores. A z-score is written as: $z = \frac{X-\mu}{\sigma}$, where X represents the particular value, μ represents the population mean, and σ represents the population standard deviation. (Literally translated, a z-score represents the number of standard deviations a value is above, or below, the mean.)For instance, if we wanted to find the area under the curve between 105 and 115 for a normal distribution with a mean of 100 and a standard deviation of 10, we would first convert our values to those appropriate for a normal distribution, $\frac{105-100}{10} = 0.5$ and $\frac{115-100}{10} = 1.5$, and then look up 1.5 and 0.5 in the table and subtract the results.

# of students	SPANISH	FRENCH	GERMAN
MALE	30	20	25
FEMALE	26	28	21

Two-Way Frequency Table

A two-way frequency table is a table that shows the number of data points falling into each combination of two categories in the form of a table, with one category on each axis. Creating a two-way frequency table is simply a matter of drawing a table with each axis labeled with the possibilities for the corresponding category, and then filling in the numbers in the appropriate cells. For instance, suppose you're told that at a given school, 30 male students take Spanish, 20 take French, and 25 German, while 26 female students take Spanish, 28 French, and 21 German. These data can be represented by the following two-way frequency table:

		Category #1 (Color)			TOTALS
		yellow	grey	black	
Category #2 (Vehicle)	truck	2	5	20	27
	car	2	25	3	30
	motorcycle	15	4	3	22
TOTALS		19	34	26	79

Joint and Marginal Frequencies

The joint frequency, $P(X,Y)$, is the probability of belonging to two specific subcategories. It's also known as a joint probability or joint relative frequency. The marginal frequency, $P(X)$, is the probability of belonging in a specific subcategory. It's also known as a marginal probability or marginal relative frequency. For example,

The joint frequency that a vehicle is a yellow truck is equal to $\frac{2}{79}$, where 2 is the number in the table cell corresponding to both "truck" and "yellow" and 79 is the total number of all vehicles of all colors.

The marginal frequency that a vehicle is a truck is equal to the sum of the probabilities of its being a yellow truck, a grey truck, or a black truck, $P(truck) = \frac{2}{79} + \frac{5}{79} + \frac{20}{79} = \frac{27}{79}$.

Conditional Frequency

The conditional frequency, $P(X|Y)$, is the probability that X is a certain value on the condition that Y is a certain value. It's also known as a conditional probability or conditional relative frequency. The conditional frequency is related to the joint frequency and the marginal frequency by the relation $P(X,Y) = P(X|Y)P(Y)$; thus knowing the joint and marginal frequencies we can find the conditional frequency. For example, consider the two-way frequency table below.

		Category #1 (Color)			TOTALS
		yellow	grey	black	
Category #2 (Vehicle)	truck	2	5	20	27
	car	2	25	3	30
	motorcycle	15	4	3	22
TOTALS		19	34	26	79

The joint frequency that a vehicle is a yellow truck is equal to $\frac{2}{79}$, and the marginal frequency that a vehicle is a truck is $\frac{27}{79}$. Therefore, the conditional frequency that a vehicle is yellow on the condition that it is a truck (the probability that a given truck in the data set is yellow) is equal to $\frac{2}{79} \div \frac{27}{79} = \frac{2}{27}$.

SCATTER PLOTS

REPRESENTING DATA ON TWO QUANTITATIVE VARIABLES

Data sets on two quantitative variables can be represented on a scatter plot by plotting each data point as a separate point on the chart. One axis of the chart corresponds to each variable, and then the data points are plotted accordingly in the appropriate positions on each axis. For instance, if 50 widgets cost \$500, 100 widgets cost \$800, and 200 widgets cost \$1500, this can be plotted on a scatter plot as follows:

Normally, of course, a scatter plot would be used for larger data sets than this. A scatter plot is useful for seeing the relationship between the variables; one can often see at a glance if the variables have a linear or other simple relationship.

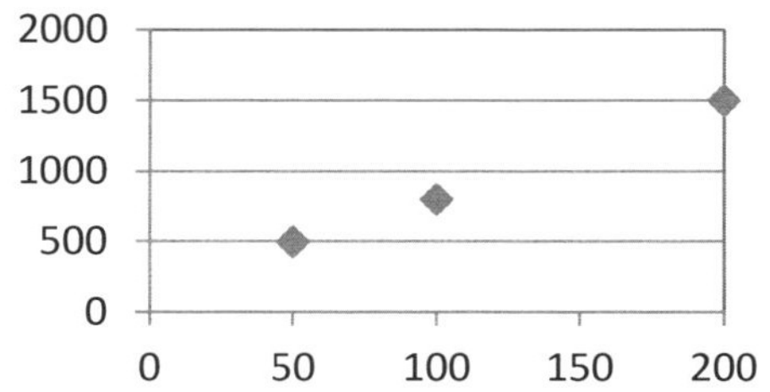

DETERMINING THE FUNCTION THAT FITS THE DATA FOR TWO VARIABLES

It's often possible from examining a scatter plot to see roughly what relationship exists between the data by estimating what kind of smooth curve would best fit the data points. Linear relationships are particularly easy to spot—do the data points look like they're roughly arranged in a straight line?—but other functions may also fit the data. Actually finding the parameters of the function for the most precise fit is a more complex matter, though most calculators and spreadsheets have the capability of performing the necessary calculations. Qualitatively judging what kind of function fits the data, however, is simpler. Below are examples of scatter plots fit to quadratic and exponential functions, respectively:

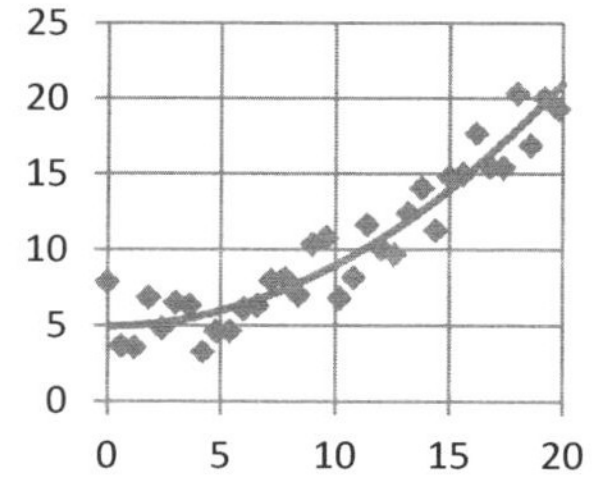

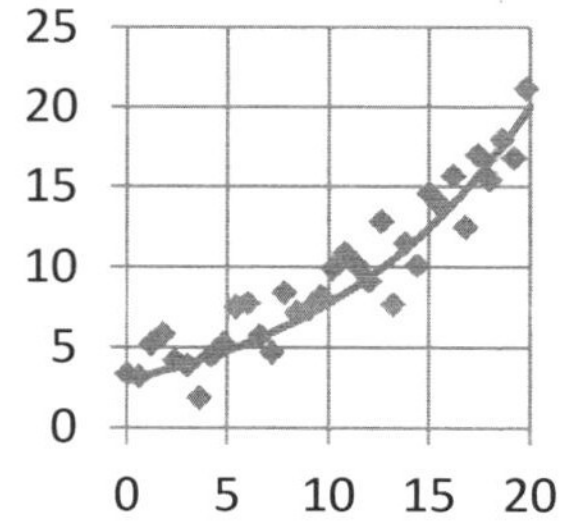

USING A LINEAR FUNCTION TO SOLVE PROBLEMS IN THE CONTEXT OF THE DATA

Once you have a function fitted to the scatter plot, you can use the equation of that function to solve problems regarding the data. For instance, suppose the scatter plot on the left below represents the number of bacteria that grow in a petri dish after one week, in millions, versus the amount of a certain nutrient added, in milligrams. The graph on the right shows a linear function fitted to the data.

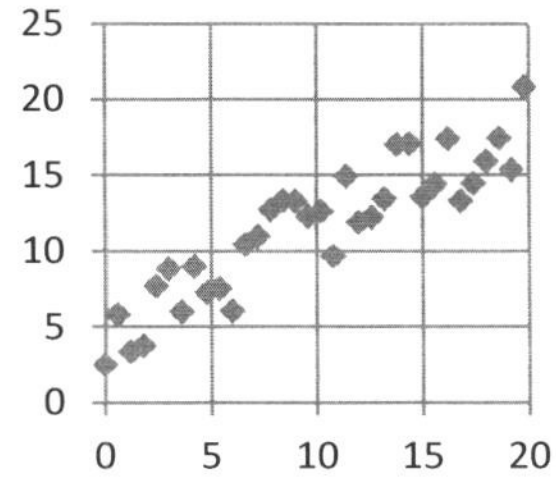

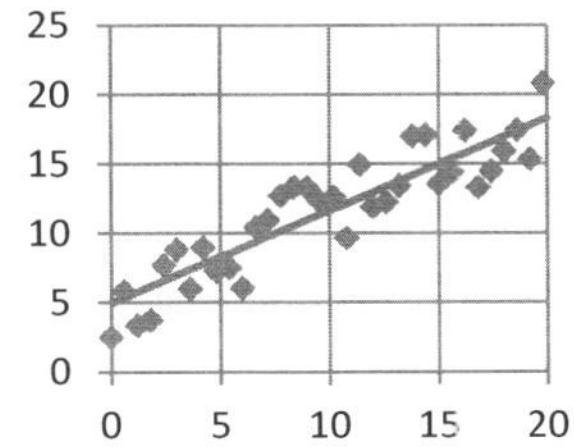

The equation of the line of best fit is $y = \frac{2}{3}x + 5$, where y is the bacteria count in millions and x is the amount of nutrient added, in milligrams. We can then use this function to solve problems; for instance, if we want to estimate how many bacteria to expect if we add 30 milligrams of nutrient, we solve $y = \frac{2}{3}(30) + 5 = 25$, or about 25 million bacteria.

Using a Quadratic Function to Solve Problems in the Context of the Data

Once you have a function fitted to the scatter plot, you can use the equation of that function to solve problems regarding the data. For instance, suppose the scatter plot on the left below represents the area covered by a certain patch of mold, in square centimeters, versus the time in days. The graph on the right shows a quadratic function fitted to the data.

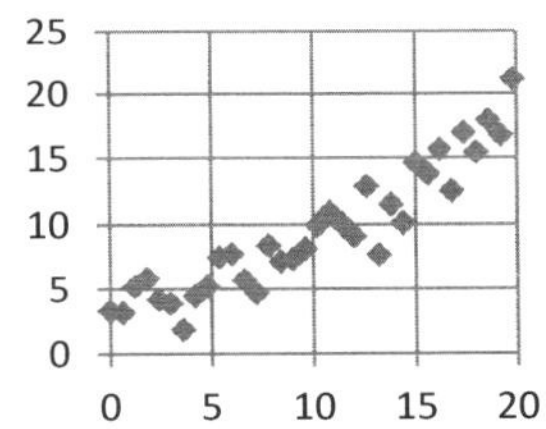

The equation of the quadratic trendline is $y = \frac{1}{25}x^2 + 5$, where x is the time in days and y is the area in square centimeters. We can then use this function to solve problems; for instance, if we want to estimate how much area to expect the mold to cover after 30 days, we solve $y = \frac{1}{25}(30^2) + 5 = 41$, or 41 square centimeters.

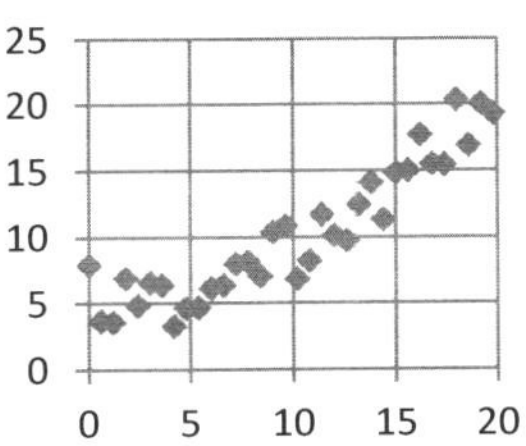

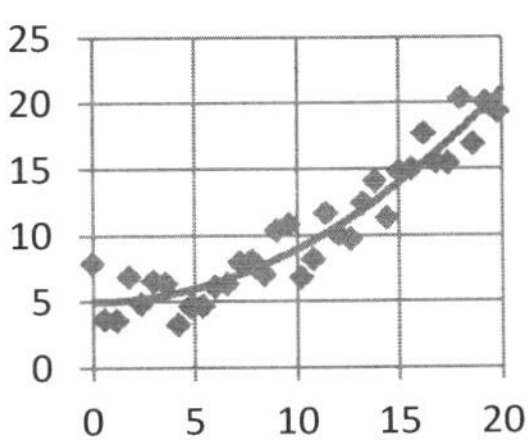

Using an Exponential Function to Solve Problems in the Context of the Data

Once you have a function fitted to the scatter plot, you can use the equation of that function to solve problems regarding the data. For instance, suppose the scatter plot on the left below represents the total assets of a certain company in millions of dollars, versus the time in years since 1990. The graph on the right shows an exponential function fitted to the data.

The equation of the trendline is $y = 3e^{0.095t}$, where t is the time in years and y is the company's assets, in millions of dollars. We can then use this function to solve problems; for instance, if we want to estimate the company's projected assets in the year 2020 (30 years after 1990), we solve $y = 3e^{0.095 \cdot 30} \approx 51.9$, or 51.9 million dollars.

RESIDUALS OF A FIT TO A DATA SET

The residuals of a fit to a data set are the differences between the observed values of the data and the predicted values based on the fit. For example, consider the data set {(6,6), (8,12), (10,20), (12,30)}, shown in the scatter plot on the left below A linear fit to the graph is shown on the right:

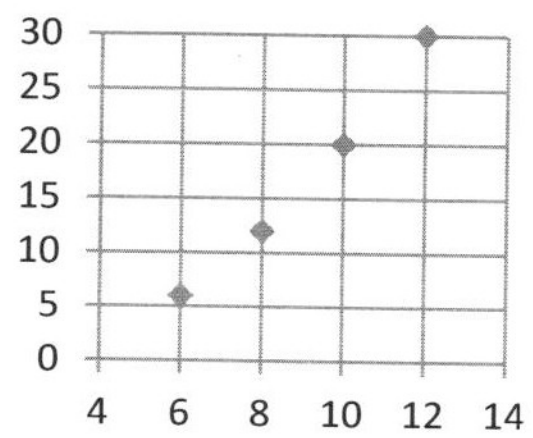

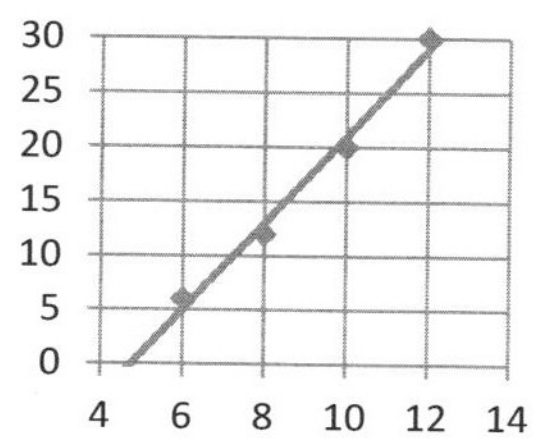

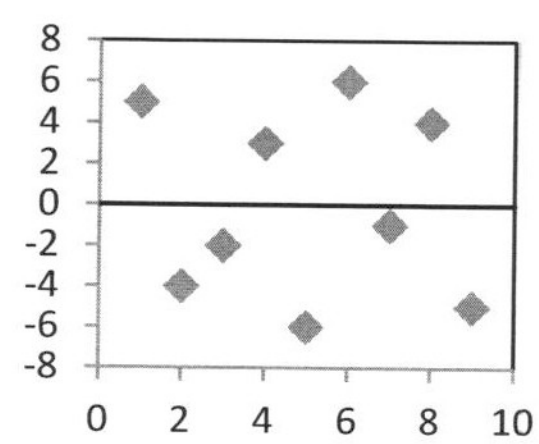

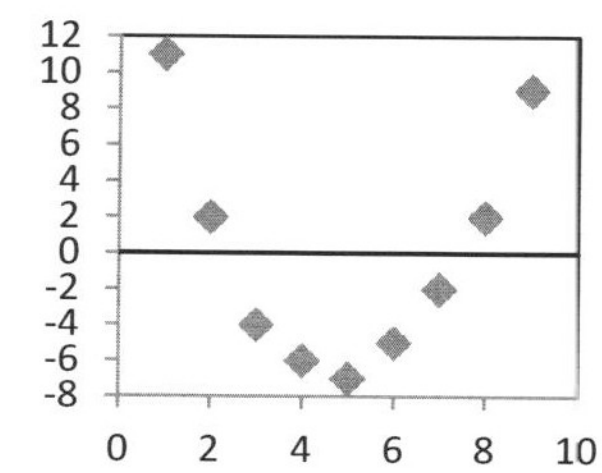

The line of best fit has the equation$y = 4x - 19$. The predicted value for x = 6 is then $4(6) - 19 = 5$, and the residual at x = 6 is $6 - 5 = 1$. At x = 8, the predicted value is $4(8) - 19 = 13$, and the residual at x = 8 is $12 - 13 = -1$. Similarly, the residuals at x = 10 and x = 12 are –1 and 1, respectively.

Note that these residuals add to 0 $(1 + (-1) + (-1) + 1 = 0)$. For the best fit curve, this is always the case; the sum and mean of the residuals will always be zero.

RESIDUAL PLOT

A residual plot is simply the plot of the residuals of a fit to a data set versus the independent variable (the x coordinates of the points). If the residual plot looks random—if the residuals seem to bear no relation to the independent variable—then the function fit to the model was probably a good choice. If the residual plot shows a definite pattern, then the data is probably better suited to a different kind of curve. Consider, for instance, the two residual plots shown below, both corresponding to linear fits. The plot on the left shows no obvious pattern, indicating that the linear fit in question is appropriate to the data. The plot on the left, however, shows a pronounced U shape, indicating that a nonlinear function would be a better fit to the data.

CALCULATING A FIT TO A LINEAR FUNCTION

WITHOUT TECHNOLOGY

Although it's complicated to calculate an exact fit to a linear function without the aid of technology, finding an approximate fit is feasible. To produce such a fit, first draw a line that seems to follow the data points as closely as possible, probably with about as many data points above the line as below it. Then find the equation for this linear fit the same way as you would find the equation of any graphed line: choose two points on the line, (x_1, y_1) and (x_2, y_2), and then find the slope of the line as $m = \frac{y_2 - y_1}{x_2 - x_1}$. Then use this value of m and one of the known points in the equation $y = mx + b$ to solve for b (it doesn't matter which of the two points you use).

Because this method involves some eyeballing and approximation, it will generally not be completely optimal, and two people may get slightly different results. However, this approximate result will typically be good enough to find estimated solutions to problems.

Using Technology

Most modern calculators and spreadsheets have the functionality to calculate linear fits to data sets. On a TI-84, you can find the fit as follows: First, press "STAT", select "EDIT", and enter your data into the table, with the independent variable in the L1 column and the dependent in the L2. Then press "STAT" again, select "CALC", and select "LinReg(ax+b)". The screen will display the coefficients a and b of the linear fit.

In recent versions of Microsoft Excel, you can put the independent and dependent variables in adjacent columns, select all the data, and insert a scatter plot. Then right-click on one of the data points of the graph and select "Add Trendline". Select "Linear", and be sure to check "Display Equation on chart" before clicking OK. The linear plot should be displayed on the graph, along with the corresponding equation of the line.

Slope

In the case of a linear fit to a data set, of the form $y = mx + b$, the slope m corresponds to the rate of change of the dependent variable y with respect to the independent variable x. This can often be expressed in a form similar to "y per x". For instance, if the data represents the distance of an object from some point as a function of time, with the distance as y and the time as x, then the slope of the linear model represents the change in distance with respect to the corresponding change in time—i.e., the velocity. If the data represents the cost to produce various quantities of products, then the slope of the linear model is the change in the cost with respect to the quantity of products produced—i.e., the production cost per unit of product.

Intercept

In the case of a linear fit to a data set, of the form $y = mx + b$, the intercept b corresponds to the value of the dependent variable y when the independent variable x is equal to zero. This often can be expressed as the "initial value" of the variable, or as its offset. For instance, if the data represents the distance of an object from some point as a function of time, with the distance as y and the time as x, then the intercept of the linear model represents the object's distance at time zero—i.e., its initial distance. If the data represents the cost to produce various quantities of products, then the intercept of the linear model is the cost when no units are being produced—in other words, the overhead cost involved in the production.

Correlation Coefficient of a Linear Fit

The correlation coefficient of a linear fit is a number that expresses how closely the linear fit approximates the function. The coefficient is negative if the line of best fit has a negative slope—the dependent variable decreases as the independent variable increases, and positive if the best fit has a positive slope—the dependent variable increases as the independent variable increases. The linear fit is most closely correlated to the data if the correlation coefficient is equal to ±1; this means that all the data points lie exactly on the best fit line. If the correlation coefficient is equal to 0, then the data are completely uncorrelated; there is no relationship between the dependent and the independent variable. More often, the correlation coefficient is somewhere in between 0 and 1, indicating a fit that reveals some correlation between the variables.

Most modern calculators and spreadsheets have the functionality to calculate the correlation coefficients of linear fits to data sets. On a TI-84, you can find the fit as follows: First, press "STAT", select "EDIT", and enter your data into the table, with the independent variable in the L1 column and the dependent in the L2. Then press "STAT" again, select "CALC", and select "LinReg(ax+b)". In addition to the coefficients a and b of the linear fit, the screen will display the correlation coefficient, r.

In recent versions of Microsoft Excel, you can put the independent and dependent variables in adjacent columns, then in another cell type "CORREL(". Select the cells containing the independent variables, then type a comma, select the cells containing the dependent variables and type a closing parenthesis. The cell should contain something like "CORREL(A1:A10,B1:B10)", though the letters and numbers may differ. Press ENTER and the cell will show the correlation coefficient of a linear fit to the data.

CORRELATION

Two variables are correlated if some nonrandom relationship exists between them: if a change in one variable tends to correspond to a change in the other. While many relationships may exist between variables—they may have a quadratic relationship, or an exponential, for example—"correlation" often refers to linear correlation, the existence of a linear relationship $y = mx + b$ between the variables. The data points need not perfectly follow that line; if they do, they are perfectly correlated, but data that approximately follow close to a line may still be said to be correlated. The degree of correlation may be quantified by a value called the correlation coefficient.

Note that correlation is not the same thing as causation; just because two variables are correlated does not necessarily mean that one is the cause of the other.

CORRELATION AND CAUSATION

Two variables are correlated if a change in one variable is accompanied by a change in the other, especially if the two variables have a linear relationship. Causation exists if one variable directly depends on the other; a change in one variable causes a change in the other. Note that correlation does not imply causation. If two variables x and y are correlated, it could be because x causes y. However, it could also be that the apparent correlation is coincidental, or that y causes x, or it could be that both x and y are influenced by a third variable separate from both x and y.

CORRELATION NOT LINKED BY CAUSATION

Many examples could be constructed of data sets with variables that are correlated but in which one variable is not the cause of the other. One way to construct such a data set is to consider two variables that might both be affected by a third. For example, suppose a survey at a given school indicates that students' shoe sizes are correlated with their spelling; the students with larger shoe sizes tend to have better spelling. Clearly it would seem strange that larger feet would lead to better spelling. In fact, there's a third variable at play here: older students would tend to have larger feet, and likewise older students will tend to have better spelling. The larger shoe sizes don't cause the better spelling, nor does the better spelling cause the larger shoe sizes; rather, both are caused by a third variable that wasn't measured.

STATISTICS

Statistics may be descriptive or inferential, in nature. Descriptive statistics does not involve inferences and includes measures of center and spread, frequencies, and percentages. Inferential statistics involves the process of making inferences about large populations based on the characteristics of random samples from the population. This field of statistics is useful because, for large populations, it may be impractical or impossible to measure the characteristics of each element of the population and determine the distributions of various properties exactly. By applying statistical methods, it's possible to make reasonable inferences about the distributions of a variable, or variables, throughout large populations by only examining a relatively small part of it. Of course, it's important that this be a "representative sample", that is, one in which the distribution of the variables of interest is similar to its distribution in the entire population. For this reason, it's desirable to create random samples, rather than choosing a set of similar samples.

Consistency of a Model

One way to determine whether a model is consistent with a given data-generating process is to repeatedly run the process, or a simulation of the process, and examine the distribution of the results, comparing them with the distribution of the model. If the distribution of the experimental or simulated results seem to have a similar shape, center, and spread to that predicted by the model, then the model is consistent with the process. If there is a significant discrepancy, then the model is not consistent with the results.

For instance, suppose you want to test whether a die is fair. This implies a model in which each number is equally likely to come up on the die. You could test this by rolling the die repeatedly and checking that indeed the numbers do come up approximately the same number of times. The distribution of the results from this experiment, when plotted, should be approximately normal.

Trials

When testing the consistency of a model, it's important to use a large number of trials in order to minimize the results of small random fluctuations. For small numbers of data points, small errors or coincidental runs of data may throw off the results and lead to false conclusions. With large amounts of data, significant deviations from the expected results become increasingly unlikely, and a match or mismatch with a model becomes increasingly likely to be significant.

For instance, suppose you want to test whether a die is fair. If you roll the die three times and it comes up 1 every time, this may seem to be inconsistent with the model, and may lead to the conclusion that the die is unfair. However, for such a small number of trials, it's not too unlikely that this anomalous result occurred by chance. If you roll the die a thousand times and it comes up 1 five hundred times, then you can be much more confident that the die is unfair.

Sample Survey, Experiment, and Observational Study

Sample surveys, experiments, and observations are all ways of measuring variables within a population. A sample survey simply polls a sample of the population and asks a set of Likert-scale and/or open-ended questions. An experiment often seeks to find a relationship between variables. In this type of study, treatment and control groups are employed Experiments are especially useful to test specific models, to see whether the experimental results match those predicted by the model. There may be circumstances, however, when it's not possible to directly control the variables of interest, either for practical or for moral reasons. In this case, an observational study may be conducted, in which one observes an individual or group, within a population, without having any direct impact on the included variables.

Sample Survey

A sample survey consists of obtaining data from a subset of the population. The data are a measure of a variable. The entire population does not need to be included in the study, but to obtain the most accurate measure of the mean requires sampling without bias, since bias will shift the estimated value of the mean.

When the variable being measured has different centers and/or spreads for its subsets, a more accurate estimate of the mean and the variance of the population can be obtained by a method called stratification, wherein the population is divided into more homogeneous subsets and the sample selected contains the same proportion of each subset as contained in the population.

The number of data points in the sample must only be large enough to obtain estimates of the parameters within pre-determined confidence intervals.

Estimating a population mean or proportion: In a well-conducted sample survey, the mean of a given variable within the surveyed sample should be approximately equal to the mean of the population as a whole. The same is true of the proportion of the sample to have a given variable within a particular range: this should be approximately equal to the proportion of the population as a whole with the variable in this range.

For instance, suppose that in a sample survey of a thousand mature animals of a particular species, the sampled individuals have an average mass of 21 kilograms, and five percent of the animals in the sample have a mass of more than 30 kilograms. Assuming that this is a well-conducted survey with a suitably chosen random sample, we would then expect the mean mass of all mature animals of the species to be 21 kilograms, and five percent of all mature animals of the species to have a mass of more than 30 kilograms.

EXPERIMENT

An experiment is a procedure by which the experimenter controls one or more "controlled variables" and then records one or more "observations". In an experiment, a treatment or intervention is given. The general purpose of an experiment is to determine the causal relationship, if any, between variables.

The simplest form of an experiment involves data with two variables (X, Y), where X is the controlled variable and Y is the observation. Generally, a function is "fitted" to the data so that it predicts the value of Y when given the value of X. The function is of the form, $\hat{y}$=f(x), where $\hat{y}$ is the predicted value of the observation, Y, for a given value$x$ of the controlled variable, X. For example, suppose $\hat{y} = f(x) = 3 + 4x$. In one run of the experiment, $(x_2, y_2) = (2{,}10.5)$, so $\hat{y} = 3 + 4(2) = 11$ and the residual is "observed value" $-$ "predicted value" $= 10.5 - 11.0 = -0.5$.

OBSERVATIONAL STUDY

An observational study is a procedure in which a researcher examines/observes an individual, or group. A set of a priori questions may or may not be used in the study.

With observational studies, the researcher(s) does not interfere with the research process, by way of given treatments. The goal is for the researcher to have little or no influence on the naturally occurring process.

ROLE OF RANDOMIZATION

Randomization is important in sample surveys, experiments, and observational studies to avoid bias. In a sample survey, for instance, if the surveyor surveys only his acquaintances or some other handpicked group, he may exhibit an unconscious bias toward individuals with certain properties, distorting his results. A random selection from the population minimizes bias, though if the population comprises discrete subsets with different characteristics it may be best to randomly select individuals within each subset, the number from each subset proportional to the total size of the subset.

Similarly, in experiments and observational studies, the experimental or observational subjects should be chosen randomly, subject to the parameters of the experiment.

MARGIN OF ERROR OF A RANDOM SAMPLING

The margin of error of a random sampling is the expected discrepancy between the mean and other characteristics of the variables in the random sampling, and those in the entire population. Generally, the larger the random sample relative to the entire population, the smaller the margin of error, to an extreme of a margin of error of zero when the sample comprises the entire population.

A margin of error can be estimated through a simulation, by setting up data set similar to the expected data in the population to be sampled, and then simulating choosing a sample of that data. By repeating this procedure, each time comparing the means of the variables as extrapolated from the sample to those of the entire simulated population, one can get a good idea of the margin of error to expect.

Randomized Experiment to Compare Two Treatments

A randomized experiment can be useful to compare two treatments to see which, if either, is more effective. If one group of patients is given one treatment and one given another, then variables related to whatever effect the treatment is supposed to produce can be measured and compared between the two groups—for instance, if the treatment is intended to cure a particular disease, then if each of two groups of inflicted patients are given one of the treatments, their rates of recovery can be measured and compared.

To ensure that the treatments have any effect at all, the treated groups should also be compared against a group that is not given any treatment. In fact, to eliminate any psychosomatic effects from the patients knowing they are being treated, they should also be compared with a control group given a placebo with no actual medicinal effect.

Simulations to Determine Significance of Differences Between Parameters

The fact that the parameters of two groups in an experimental study are different does not necessarily mean that this difference is a product of the difference in the control variables. Some variation inevitably occurs, and it could be that the difference is simply the result of such statistical error. One way to test this is to use a two-sample z-test. Hypothesis testing will reveal a p-value that may be compared to an a priori level of significance. For a p-value less than the stated level of significance, the means of the two groups may be determined to be statistically different. (In this case, the null hypothesis of no difference would be rejected.) Repeated studies may be used to validate findings.

Using Data to Evaluate Reports

Reports of results of trials and experiments generally include relationships and conclusions inferred from the experimental data. However, some reports may make faulty inferences and come to conclusions not justified by the data. To evaluate a report, it helps to check the data yourself and see if the conclusions stated are really warranted by the data. There are a number of questions to look out for. Did the writer of the report assume that correlation implied causation? Did he draw definitive conclusions from an apparent result that is too small to be significant? Did he randomly choose the samples for his experiment or survey, or did he select them in a way that might lead to possible bias? These are among the common mistakes to look out for, both in drawing conclusions from a study yourself and in evaluating a report written by someone else.

Sample Space of a Statistical Experiment

The sample space of a statistical experiment is the set of all possible distinct outcomes of the experiment. If the experiment comprises the flip of a coin, for example, there are only two elements in the sample space: either the coin will come up heads, or the coin will come up tails. For the roll of a standard die, there are six possible outcomes in the sample space, one for each number that could come up on the die. When drawing a card from a full deck (without jokers), there are 52 possible outcomes.

The elements of the sample space need not all have the same probability. If you roll a weighted die that comes up 1 50% of the time, the sample space is the same as for the roll of an ordinary, unweighted die; only the probabilities of the elements in the sample space differ.

Event in a Statistical Experiment

In a statistical experiment, an event can be defined as a subset of the sample space—that is, an event is a particular collection of possible outcomes. The event may have only one element, or it may have many different elements. For instance, suppose the statistical experiment in question is the draw of a card from a deck. One event is the draw of the ace of spades—this is one possible outcome from the draw. Another could be the draw of a jack—this event comprises four possible outcomes, since there are four jacks in the deck. Still another could be the draw of a black card—twenty-six of the cards in the deck are black, so this event comprises twenty-six different outcomes.

Union and Intersection of Two Sets of Outcomes

Let A and B each be a set of elements or outcomes. Since the application is probability theory, we will use the term "outcomes".

The union (symbol ∪) of two sets is the set of elements found in Set A or Set B. For example, if A={2,3,4} and B={3,4,5}, $A \cup B = \{2, 3, 4, 5\}$. Note that the outcomes 3 and 4 appear only once in the union. For statistical events, the union is equivalent to "or"; $P(A \cup B)$ is the same thing as P(A or B).

The intersection (symbol ∩) of two sets is the set of outcomes common to both sets. For the above sets A and B, $A \cap B = \{3, 4\}$. For statistical events, the intersection is equivalent to "and"; $P(A \cap B)$ is the same thing as P(A and B).

The union and intersection commute: $A \cup B = B \cup A$and$A \cap B = B \cap A$.The union of sets is left and right distributive over intersection, $A \cup (B \cap C) = (A \cup B) \cap (A \cup C)$and $(B \cap C) \cup A = (B \cup A) \cap (C \cup A)$. Likewise, the intersection of sets is left and right distributive over union: $A \cap (B \cup C) = (A \cap B) \cup (A \cap C)$ and $(B \cup C) \cap A = (B \cap A) \cup (C \cap A)$.

Complement of a Set of Outcomes

In general, the complement of a set is the set of all possible elements not in the set. This requires first defining the set of all possible elements under consideration, known as the universe. In probability theory, however, this is already defined; the universe under consideration is the sample space. The complement of a set of outcomes is therefore the set of all outcomes in the sample space that are not in the original set.

For example, suppose that an integer is randomly chosen between 1 and 7. The sample space is then {1, 2, 3, 4, 5, 6, 7}. Consider the set of outcomes A = {1, 3, 4}. Then its complement A^C = {2, 5, 6, 7}—all the possible outcomes in the sample set that do not appear in A.

The complement of a union of two sets is the intersection of their complements, and vice versa: $(A \cup B)^C = A^C \cap B^C$, and $(A \cap B)^C = A^C \cup B^C$.

Defining Events in Terms of Other Events

Consider the draw of a card from a deck. Drawing the ace of spades is one possible event; drawing a red card is another; drawing a face card is still another. Still another event is the drawing of either the ace of spades or a red card—since both the outcomes of the two events "ace of spades" and "red card" are included, this event is the union of those two events. We can also consider drawing a red face card, i.e. a card that is both a red card and a face card. The set of outcomes in this event is the intersection of those in the two events "red card" and "face card". The drawing of a card that is not a face card is also an event, the complement of the face card event. In general, "or" indicates a union, "and" an intersection, and "not" a complement of events.

INDEPENDENT EVENTS

Two events are independent if the occurrence of one event has no effect on the probability of the other event occurring. For example, for two successive coin flips, the coin comes up heads on the first flip and the coin coming up heads on the second flip are independent events: whether the coin comes up heads or tails on the first flip has no effect on whether it comes up heads or tails on the second. However, if you draw two cards from a deck, while not replacing the first card, the probability of the first and second cards being red represents dependent events. In this case, the number of possible outcomes and the sample space, in the second event, are each decreased by 1.

PROVING INDEPENDENCE WITH THE PRODUCT OF THE PROBABILITIES

For two independent events, and only for independent events, the product of the probabilities of each event is equal to the probability that both events will occur: $P(A)P(B) = P(A \cap B)$. If we know these probabilities, then, we can use this equation to check whether the events are independent. Conversely, if we know the probabilities of two events that we know are independent, we can use this equation to find the probability that both will occur. For example, suppose you're programming a computer game in which there's a random chance of finding a certain item at the end of each level. You want the chances at each level to be independent. Suppose you find after running some trials that the chances of finding the item at the end of the first level are $\frac{1}{3}$, at the end of the second level are $\frac{1}{4}$, and at the end of both the first and the second level are $\frac{1}{12}$. Since $\left(\frac{1}{3}\right)\left(\frac{1}{4}\right) = \left(\frac{1}{12}\right)$, this proves that the events are indeed independent.

CONDITIONAL PROBABILITY

Given two events A and B, the **conditional probability** $P(A|B)$ is the probability that event A will occur, given that event B has occurred. For instance, suppose you have a jar containing two red marbles and two blue marbles, and you draw two marbles at random. Note. The first drawn marble is not replaced. Consider event A being the event that the first marble drawn is red, and event B being the event that the second marble drawn is blue. With no conditions set, both P(A) and P(B) are equal to $\frac{1}{2}$. However, if we know that the first marble drawn was red—that is, that event A occurred—then that leaves one red marble and two blue marbles in the jar. In that case, the probability that the second marble is blue given that the first marble was red—that is, P(A|B)—is equal to $\frac{2}{3}$.

		Size			
		Small	**Medium**	**Large**	**Total**
Color	**Blue**	25	40	35	100
	White	27	25	22	74
	Black	8	23	15	26
	Total	60	88	72	220

Calculating the Conditional Probability P(A|B) in Terms of the Probabilities of Events A and B and Their Union and/or Intersection

The conditional probability P(A|B) is the probability that event B will occur given that event A occurs. This cannot be calculated simply from P(A) and P(B); these probabilities alone do not give sufficient information to determine the conditional probability. It can, however, be determined given also $P(A \cap B)$, the probability that events A and B both occur. Specifically, $P(A|B) = \frac{P(A \cap B)}{P(B)}$. For instance, suppose you have a jar containing two red marbles and two blue marbles, and you draw two marbles at random. Consider event A being the event that the first marble drawn is red, and event B being the event that the second marble drawn is blue. P(A) is $\frac{1}{2}$, and $P(A \cap B)$ is $\frac{1}{3}$. (The latter may not be obvious, but may be determined by finding the product of $\frac{1}{2}$ and $\frac{2}{3}$) Therefore $P(A|B) = \frac{1/3}{1/2} = \frac{2}{3}$.

Estimating a Conditional Probability from a Two-Way Frequency Table

If we have a two-way frequency table, it is generally a straightforward matter to read off the probabilities of any two events A and B, as well as the joint probability of both events occurring, $P(A \cap B)$. We can then find the conditional probability $P(A|B)$ by calculating $P(A|B) = \frac{P(A \cap B)}{P(B)}$. We could also check whether or not events are independent by verifying whether $P(A)P(B) = P(A \cap B)$.

For example, a certain store's recent T-shirt sales:

Size / Color	Small	Medium	Large	Total
Blue	25	40	35	100
White	27	25	22	74
Black	8	23	15	46
Total	60	88	72	220

Suppose we want to find the conditional probability that a customer buys a black shirt (event A), given that the shirt he buys is size small (event B). From the table, the probability $P(B)$ that a customer buys a small shirt is $\frac{60}{220} = \frac{3}{11}$. The probability $P(A \cap B)$ that he buys a small, black shirt is $\frac{8}{220} = \frac{2}{55}$. The conditional probability $P(A|B)$ that he buys a black shirt, given that he buys a small shirt, is therefore $P(A|B) = \frac{2/55}{3/11} = \frac{2}{15}$.

Conditional Probability in Everyday Situations

Conditional probability often arises in everyday situations in, for example, estimating the risk or benefit of certain activities. The conditional probability of having a heart attack given that you exercise daily may be smaller than the overall probability of having a heart attack. The conditional probability of having lung cancer given that you are a smoker is larger than the overall probability of having lung cancer. Note that changing the order of the conditional probability changes the meaning: the conditional probability of having lung cancer given that you are a smoker is a very different thing from the probability of being a smoker given that you have lung cancer. In an extreme case, suppose that a certain rare disease is caused only by eating a certain food, but even then is unlikely. Then the conditional probability of having that disease given that you eat the dangerous food is nonzero but low, but the conditional probability of having eaten that food given that you have the disease is 100%!

Independent Events

The conditional probability P(A|B) is the probability that event B will occur given that event A occurs. If the two events are independent, we do not expect that whether or not event A occurs should have any effect on whether or not event B occurs. In other words, we expect P(A|B)=P(A).

This can be proven using the usual equations for conditional probability and the joint probability of independent events. The conditional probability $P(A|B) = \frac{P(A \cap B)}{P(B)}$. But if A and B are independent, then $P(A \cap B) = P(A)P(B)$. So $P(A|B) = \frac{P(A)P(B)}{P(B)} = P(A)$.

By similar reasoning, if A and B are independent then P(B|A)=P(B).

Independence

Determining Independence from a Two-Way Frequency Table

If we have a two-way frequency table, it is generally a straightforward matter to read off the probabilities of any two events A and B, as well as the joint probability of both events occurring, $P(A \cap B)$. We can then check whether or not the events are independent by verifying whether $P(A)P(B) = P(A \cap B)$.

For example, consider the following table, showing a certain store's recent T-shirt sales, categorized by size and color:

		Size			
		Small	**Medium**	**Large**	**Total**
Color	**Blue**	25	40	35	100
	White	27	25	22	74
	Black	8	23	15	26
	Total	60	88	72	220

Suppose we want to check whether the event A that a customer buys a blue shirt is independent of the event B that a customer buys a medium shirt. From the table, $P(A) = \frac{100}{220} = \frac{5}{11}$ and$P(B) = \frac{88}{220} = \frac{4}{10}$. Also, $P(A \cap B) = \frac{40}{220} = \frac{2}{11}$. Since $\left(\frac{5}{11}\right)\left(\frac{4}{10}\right) = \frac{20}{220} = \frac{2}{11}$, $P(A)P(B) = P(A \cap B)$ and these two events are indeed independent.

Everyday Situations

Even if we don't necessarily think of it in those terms, we often use independence of events in everyday situations. We know when playing a board game, for instance, that each roll of the die is independent, and is unaffected by previous rolls.

Sometimes, however, our intuition fails us and we tend to think of independent events as influencing each other; if we flip a coin and it comes up heads three times in a row, we may erroneously believe on some level that the next flip is more likely to come up tails because a tails is "overdue". Many gamblers make this mistake, seeing meaningful streaks in random results, and expecting certain results on the basis of previous events. Remembering what it means for events to be random helps avoid this error.

Uniform Probability Model

A uniform probability model is a model in which the probabilities of all outcomes are equally likely. For example, the roll of a fair die can be simulated by such a model; it's equally likely that any face of the die come up on top. Likewise, the draw of a card from a well-shuffled deck can be represented by a uniform probability model; it's equally likely for any card to be drawn.

One useful feature of a uniform probability model is that probabilities can be determined by simply counting outcomes. The probability of an event A occurring is equal to the total number of outcomes in event A divided by the total number of outcomes in the sample set. For example, suppose we want to know the probability of drawing a face card from a well-shuffled deck of cards (without jokers). There are fifty-two cards in the deck, twelve of which are face cards. The probability of drawing a face card from the deck is therefore $\frac{12}{52} = \frac{3}{13}$.

When plotted, a uniform distribution is rather flat, in appearance. Uniform distribution results when plotting the sampling distribution of a large number of samples, thus illustrating the Central Limit Theorem.

Finding the Conditional Probability P(A|B)

Although one way of finding the conditional probability is to calculate $P(A|B) = \frac{P(A \cap B)}{P(B)}$, this is not the only way. For a uniform probability model, it's possible to find the conditional probability even without knowing the overall probability of any event, just knowing all the outcomes included in event B. To see this, consider that the probability of event B occurring is equal to the number of outcomes in event B, which we can call X_B, divided by the total number of outcomes in the sample space, X_{TOTAL}. Similarly, the probability of events A and B both occurring is equal to the number of outcomes in the intersection of A and B, $X_{A \cap B}$, divided by the total number of outcomes in the sample space, X_{TOTAL}. So the conditional probability $P(A|B) = \frac{P(A \cap B)}{P(B)} = \frac{X_{A \cap B}/X_{TOTAL}}{X_B/X_{TOTAL}} = \frac{X_{A \cap B}}{X_B}$. In other words, we can find the conditional probability $P(A|B)$ by simply dividing the number of outcomes that belong to both A and B by the number of outcomes that belong to B: $P(A|B)$ is the fraction of B's outcomes that also belong to A.

Interpreting a Conditional Probability

A conditional probability $P(A|B)$ is the probability that event A will occur given that event B occurs. For a uniform probability model, this is equal to the fraction of B's outcomes that also belong to A. For instance, suppose we are interested in knowing the number of students of each grade at a certain school who buy their lunch at the school and who take their lunch from home. We will call the event that a randomly chosen student is in the ninth grade event A, and the event that he buys his lunch at school event B. Since the students are randomly chosen with equal probability, this is a uniform probability model. The probability that a student buys his lunch at school, given that he is in the ninth grade, is $P(A|B)$, and is simply equal to the fraction of ninth-grade students who buy their lunches at the school.

Addition Rule

The addition rule for probabilities is $P(A \cup B) = P(A) + P(B) - P(A \cap B)$. In other words, the probability that either event A or event B will occur is equal to the sum of the probabilities of each event, minus the probability that they will both occur.

For example, suppose you want to find the probability that, when you roll two dice, either their numbers will match (event A) or they will add to eight (event B). $P(A) = \frac{6}{36} = \frac{1}{6}$, since there are six possible combinations of matching dice ((1,1), (2,2), (3,3), (4,4), (5,5), or (6,6)), and thirty-six combinations total. $P(B) = \frac{5}{36}$; there are five combinations that add to 8 ((2,6), (3,5), (4,4), (5,3), or (6,2)). $P(A \cap B) = \frac{1}{36}$; there is only one combination of matching dice that add to 8 (that is, (4,4)). So $P(A \cup B) = \frac{1}{6} + \frac{5}{36} - \frac{1}{36} = \frac{5}{18}$.

Interpreting the Result in Terms of a Uniform Probability Model

The addition rule states that $P(A \cup B) = P(A) + P(B) - P(A \cap B)$. The result of the equation gives the probability that either event A or B will occur, which for a uniform probability distribution is equal to the fraction of outcomes in either A or B.

Suppose, for example, we are told that of the books on a certain shelf, $\frac{1}{3}$ are paperbacks, $\frac{1}{2}$ are fiction, and $\frac{1}{4}$ are paperback fiction books. We can call event A the event that a randomly chosen book is a paperback and event B the event that a randomly chosen book is fiction. Since we are randomly choosing the books, this is a uniform probability model. From the given information, $P(A) = \frac{1}{3}$, $P(B) = \frac{1}{2}$, and $P(A \cap B)$—the probability that a randomly chosen book is both paperback and fiction—is equal to $\frac{1}{4}$. So $P(A \cup B) = \frac{1}{3} + \frac{1}{2} - \frac{1}{4} = \frac{7}{12}$. This indicates in this case that $\frac{7}{12}$ of the books are either paperback or fiction.

General Multiplication Rule

The general multiplication rule for probabilities states that $P(A \cap B) = P(A)P(B|A) = P(B)P(A|B)$. In other words, the probability that events A and B will both occur is equal to the probability that event A will occur times the probability that event B will occur on the condition that A occurs; or, equivalently, it is equal to the to the probability that event B will occur times the probability that event A will occur on the condition that B occurs.

For example, suppose we're told that a certain cat is in a certain yard half the time, and when it is in the yard there is a $\frac{1}{3}$ chance that a second cat will be with it. We will call the event of the first cat being in the yard A, and the event of the second cat being in the yard B. Then from the given information, $P(A) = \frac{1}{2}$ and $P(B|A) = \frac{1}{3}$. The probability that both cats are in the yard is then $P(A \cap B) = P(A)P(B|A) = \left(\frac{1}{2}\right)\left(\frac{1}{3}\right) = \frac{1}{6}$.

Interpreting the Result in Terms of a Uniform Probability Model

The general multiplication rule states that $P(A \cap B) = P(A)P(B|A) = P(B)P(A|B)$. The result of this rule is the probability that events A and B will both occur, which for a uniform probability model is equal to the fraction of outcomes that are in both events A and B.

For example, suppose we are told that $\frac{1}{4}$ of the insects in a certain woman's insect collection are butterflies, and that $\frac{1}{5}$ of those butterflies are blue. We can call event A the event that a randomly chosen insect from the collection is a butterfly and event B the event that it is blue. So $P(A) = \frac{1}{4}$, and $P(B|A) = \frac{1}{5}$—the fraction of butterflies in the collection that are blue is equal to the probability that a randomly chosen insect in the collection is blue, given that it is a butterfly. Then by the multiplication rule, $P(A \cap B) = P(A)P(B|A) = \left(\frac{1}{4}\right)\left(\frac{1}{5}\right) = \frac{1}{20}$. So the probability that a randomly selected insect from the collection is both blue and a butterfly is $\frac{1}{20}$; $\frac{1}{20}$ of the insects are blue butterflies.

PERMUTATIONS AND COMBINATIONS

Both combinations and permutations are used when it is required to determine the number of ways that k elements can be chosen (without repetition) out of a larger set of n elements. In general, a permutation is used when the order of the elements matters, and a combination when it doesn't. For instance, if we wanted to calculate how many different five-letter sequences are possible (with no letters repeated), the order matters, so we would want the permutation of five elements out of the twenty-six-letter alphabet, $_{26}P_5$(which is equal to $\frac{26!}{(26-5)!} = 7{,}893{,}600$). If we wanted to calculate how many different hands of four cards we could draw from a deck of 52 cards, the order doesn't matter, so we would want the combination of four elements out of the fifty-two, $_{52}C_4$ (which is equal to $\frac{52!}{4!(52-4)!} = 270{,}725$.

PERMUTATION

A permutation is a way of selecting k elements out of a larger group of n elements, in which the order matters: (1, 2, 3) is considered a different permutation from (2, 3, 1). The number of permutations of k elements out of n can be written as $_nP_k$, and is equal to $\frac{n!}{(n-k)!}$. When k is small, this is more easily calculated as $(n)(n-1)\dots(n-k+1)$. For instance, there are six possible ways to choose two (ordered) elements out of three: (1, 2), (2, 1), (1, 3), (3, 1), (2, 3), (3, 2). And, accordingly, $_3P_2 = \frac{3!}{(3-2)!} = \frac{6}{1} = 6$. The number of ways to choose five (ordered) elements out of a hundred is much harder to enumerate, but not hard to calculate: $_{100}P_5 = \frac{100!}{(100-5)!} = (100)(99)(98)(97)(96) = 9{,}034{,}502{,}400$. The number of ways to choose 95 (ordered) elements out of 100 is much larger, equal to $_{100}P_{95} = \frac{100!}{(100-95)!} = \frac{100!}{5!}$, which turns out to be a number with more than 150 digits.

COMBINATION

A combination is a way of selecting k elements out of a larger group of n elements, in which the order doesn't matter: (1, 2, 3) and (2, 3, 1) are considered the same combination. The number of permutations of k elements out of n can be written as $_nC_k$, and is equal to $\frac{n!}{k!(n-k)!}$. For instance, there are three possible ways to choose two (unordered) elements out of three: (1, 2), (2, 3), or (1, 3). And, accordingly, $_3C_2 = \frac{3!}{2!(3-2)!} = \frac{6}{2\cdot 1} = 3$. The number of ways to choose five (unordered) elements out of a hundred is much harder to enumerate, but not hard to calculate: $_{100}C_5 = \frac{100!}{5!(100-5)!} = \frac{100!}{5!\cdot 95!} = 75{,}287{,}520$. Similarly, the number of ways to choose 95 (unordered) elements out of 100 is equal to $_{100}C_{95} = \frac{100!}{95!(100-95)!} = \frac{100!}{95!\cdot 5!} = 75{,}287{,}520$. This illustrates a general rule: $_nC_k = {_nC_{n-k}}$, for any positive integer values of n and k such that $k \leq n$.

CALCULATING THE NUMBER OF POSSIBLE SEQUENCES OF SIX LETTERS, WITH NO LETTERS REPEATED

The total number of possible sequences of six letters, with no letters repeated, is equal to the number of permutations of 6 elements out of 26, or $_{26}P_6$. This can be calculated as $_{26}P_6 = \frac{26!}{(26-6)!} = \frac{26!}{20!} = 165{,}765{,}600$. There are almost 166 million possible sequences of six letters, with no letters repeated.

If repetitions are allowed, then the permutation formula no longer applies. Rather, now each letter can be treated as an independent event, since choosing one letter has no bearing on the possible choices for the others. Since there are 26 possible choices for the first letter, 26 for the second, 26 for the third, and so on, the total number of combinations is $26 \cdot 26 \cdot 26 \cdot 26 \cdot 26 \cdot 26 = 26^6 =$ 308,915,776.

Probability Problem Solved with a Permutation

A permutation can be used to solve a problem that involves finding the probability of one or more particular ordered arrangements of elements occurring. The permutation can be used to find the total number of outcomes in the sample space, and then the probability is the number of outcomes in the event in question divided by that number. For example, suppose you had twelve tiles labeled with the numbers one through twelve, and you wanted to find the probability that, if you drew three tiles at random, they would be consecutive numbers in ascending order. There are ten arrangements that fit those criteria: (1, 2, 3), (2, 3, 4), (3, 4, 5), (4, 5, 6), (5, 6, 7), (6, 7, 8), (7, 8, 9), (8, 9, 10), (9, 10, 11), and (10, 11, 12). The total number of possible outcomes is ${}_{12}P_3 = \frac{12!}{(12-3)!} =$ 1320. The probability that they will fulfill the criteria is then $\frac{10}{1320} = \frac{1}{132}$.

Probability Problem That Can Be Solved with a Combination

A combination can be used to solve a problem that involves finding the probability of one or more particular unordered arrangements of elements occurring. The combination can be used to find the total number of outcomes in the sample space, and then the probability is the number of outcomes in the event in question divided by that number.

For example, suppose you wanted to find the probability of drawing a royal flush (a ten, ace, jack, queen, and king all of the same suit) from a standard deck of 52 cards. There are four suits in a standard deck, thus four possible royal flushes. Since the order of the cards doesn't matter, the total number of possible outcomes—sets of five cards that can be drawn—is equal to ${}_{52}C_5 = \frac{52!}{5!(52-5)!} = 2{,}598{,}960$. The probability of drawing a royal flush is therefore $\frac{4}{2{,}598{,}960} = \frac{1}{649{,}740}$.

Deriving Formulas for Permutations and Combinations

The number of permutations of k elements out of n, ${}_nP_k$, is equal to the number of ways we can choose k elements, in order, from a larger set of n. The first element can be any of the n elements of the larger set. Once we've chosen it, however, there are only $n - 1$ elements left to choose for the second element, then $n - 2$ for the third, and so on down to $n - k + 1$ for the kth element. The total number of possibilities is then the product of all of these numbers, $n(n-1)(n-2)\ldots(n-k+1)$, which is equal to $\frac{n(n-1)(n-2)\ldots(n-k+1)(n-k)(n-k-1)\ldots(2)(1)}{(n-k)(n-k-1)\ldots(2)(1)} = \frac{n!}{(n-k)!}$.

The number of combinations of k elements out of n, ${}_nC_k$ is equal to the number of ways we can choose k elements out of a larger set of n, disregarding the order. This is similar to ${}_nP_k$ except that the order doesn't matter. Essentially, then, we can get ${}_nP_k$ from ${}_nC_k$ by multiplying by the number of possible orders of the k elements, which is ${}_kP_k = k!$. So ${}_nP_k = {}_nC_k \cdot k!$; solving for ${}_nC_k$ we get ${}_nC_k = \frac{{}_nP_k}{k!} = \frac{n!}{k!(n-k)!}$.

Defining a Random Variable for a Quantity of Interest

In order to graph a random variable and analyze its distribution, it's helpful to quantify it, to assign a numerical value to each possible event. This is more useful when there is some number that naturally corresponds to the event, rather than just arbitrary assignments. If the random variable represents the colors of the cars passing by, there's little gained by defining 1 to correspond to blue, 2 to red, and so on. However, if you want to quantify a student's attendance in a class, for example, one can derive a suitable number by, perhaps, adding together the number of days the student has been absent plus one fourth the number of days he has been tardy. This then gives a numerical value that can be graphed and otherwise analyzed.

PROBABILITY DISTRIBUTION

A probability distribution is a set of values of a random variable, with a probability assigned to each one. All possible values of the random variable should be represented in the distribution, and the sum of all their probabilities must equal 1, or 100%. For instance, suppose our random variable represents the number of tails that come up if we flip a coin five times. The probability that no tails will come up is $\frac{1}{2^5} = \frac{1}{32}$, the probability of one tails is $\frac{5}{32}$, the probability of two or three tails is each $\frac{5}{16}$, the probability of four tails is $\frac{5}{32}$, and the probability that the coin will come up heads all five times is $\frac{1}{32}$. Note that as required, the sum of all these probabilities is equal to 1: $\frac{1}{32} + \frac{5}{32} + \frac{5}{16} + \frac{5}{16} + \frac{5}{32} + \frac{1}{32} = \frac{1+5+10+10+5+1}{32} = \frac{32}{32} = 1$.

GRAPHING

A probability distribution of a random variable may be graphed similarly to a data set. You can plot the value of the variable on the x axis and the probability corresponding to that outcome on the y axis. Since the probabilities add to 1, the individual probabilities are likely to be small, and the axis should be scaled accordingly.

For example, consider the random variable corresponding to the sum of the numbers on three fair dice. The probability that this sum will be 3 is $\frac{1}{216} \approx 0.0046$. The probability that the dice will sum to 4 is $\frac{3}{216} \approx 0.014$, and so on; the largest probabilities arise in the center, where we reach a probability of $\frac{27}{216} = 0.125$ that the sum of the dice will be 10, and the same probability for 11. If we plot the probabilities of each possible outcome on a graph, we get the following:

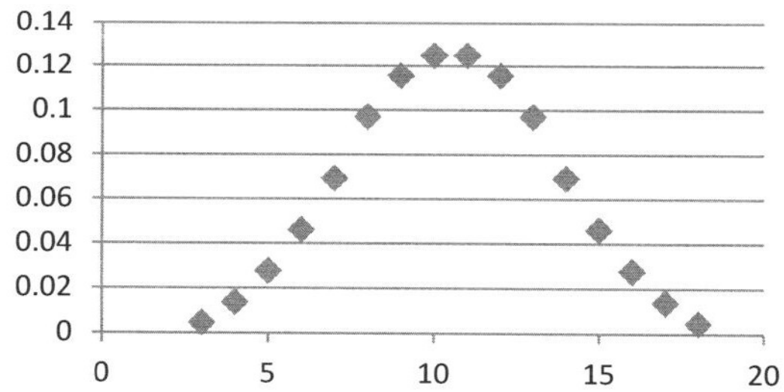

MEAN

The mean of a probability distribution is synonymous with the expected value of a random variable. It can be calculated by summing all possible values of the variable weighted by their respective probabilities, $E(X) = \sum_{i=1}^{N} x_i P(x_i)$. For a uniform distribution, with all probabilities the same, $P(x_i) = \frac{1}{N}$ for all I, and this reduces to $(X) = \frac{1}{N}\sum_{i=1}^{N} x_i$, the more familiar formula for the mean of a data set. The more general formula for the mean of a probability distribution, however, holds for arbitrary distributions that may not be uniform.

The mean serves a similar role to the mean of a data set; it provides a measure of central tendency, and a rough idea of what a typical data point looks like. However, it also suffers from some of the same limitations as the mean of a data set, and does not by itself provide a complete picture of the distribution.

RANDOM VARIABLE IN A SAMPLE SPACE IN WHICH THEORETICAL PROBABILITIES CAN BE CALCULATED

For a random variable in a sample space in which theoretical probabilities can be calculated, developing a probability distribution is just a matter of calculating the theoretical probabilities for each value of the variable, and then assigning each one to the appropriate variable.

For example, suppose there is a multiple choice test with five questions, each with four choices. The probability of a student guessing all five questions correctly is $\left(\frac{1}{4}\right)^5 \approx 0.00098$. The probability of his guessing four questions correctly is $\left(\frac{1}{4}\right)^4 \left(\frac{3}{4}\right) {}_5C_4 \approx 0.01465$. The chances of his guessing three questions correctly is $\left(\frac{1}{4}\right)^3 \left(\frac{3}{4}\right)^2 {}_5C_3 \approx 0.08789$, of guessing two questions correctly $\left(\frac{1}{4}\right)^2 \left(\frac{3}{4}\right)^3 {}_5C_2 \approx 0.26367$, of guessing one question $\left(\frac{1}{4}\right) \left(\frac{3}{4}\right)^4 {}_5C_1 \approx 0.39551$, and of his guessing all the questions incorrectly $\left(\frac{3}{4}\right)^5 \approx 0.23731$. If we define our random variable to be the number of questions that a student guesses correctly, then matching these probabilities to the corresponding values of the random variable gives us an appropriate probability distribution.

PROBABILITY DISTRIBUTION FOR A RANDOM VARIABLE DEFINED FOR A SAMPLE SPACE IN WHICH PROBABILITIES ARE ASSIGNED EMPIRICALLY

In developing a probability distribution for a sample space in which probabilities are assigned empirically, most of the work is done for you. You just have to define your random variable, and then assign to each value of the random variable the appropriate probability.

For example, suppose that you find data on pet ownership in a particular neighborhood claiming that 22% of the neighborhood's households own no pets, 30% own one pet, 19% own two, 13% own three pets, 9% own four pets, 4% own five pets, 3% own six pets, and 1% own seven pets. A natural choice for a random variable is the number of pets owned by a particular household in the neighborhood. The corresponding probabilities are therefore just the quoted percentages, since these are equal to the probabilities that a randomly chosen household in the neighborhood will have that number of pets.

EXPECTED VALUE OF A RANDOM VARIABLE

The expected value of a random variable is the sum of the possible values of the variable weighted by the probability of that value occurring. Mathematically, it can be expressed as $E(X) = \sum_{i=1}^{N} x_i P(x_i)$.

For instance, suppose you have a weighted die that has a probability $\frac{1}{3}$ of coming up 6, a probability $\frac{1}{6}$ of coming up 5, and a probability $\frac{1}{8}$ of coming up 1, 2, 3, or 4. The expected value of the die's result is therefore $6\left(\frac{1}{3}\right) + 5\left(\frac{1}{6}\right) + 4\left(\frac{1}{8}\right) + 3\left(\frac{1}{8}\right) + 2\left(\frac{1}{8}\right) + 1\left(\frac{1}{8}\right) = 2 + \frac{5}{6} + \frac{1}{2} + \frac{3}{8} + \frac{1}{4} + \frac{1}{8} = \frac{48+20+12+9+6+3}{24} = \frac{98}{24} = \frac{49}{12} \approx 4.08$.

Note that, as seen in this example, the expected value is not necessarily a value that you would actually expect to see; the weighted die will never show 4.08. Rather, it plays a role similar to the mean of a data set, providing a measure of central tendency of the variable's distribution.

For a Sample Space in Which Theoretical Probabilities Can Be Calculated

The first step in finding the expected value of a random variable is to calculate the probabilities for each value. After that, we can find the expected value by summing all the possible values of the random variable, weighted by the respective probabilities, $E(X) = \sum_{i=1}^{N} x_i P(x_i)$.

For example, suppose there is a multiple choice test with five questions, each with four choices. We can set our random variable to be the number of questions a student guesses correctly. Without going into details of the calculations, the probability of a student guessing all five questions correctly is about 0.00098. The probability of his guessing four questions correctly is about 0.01465, three 0.08789, two 0.26367, one 0.39551, none 0.23731. The expected value for our random variable is therefore $E(X) = 5(0.00098) + 4(0.01465) + 3(0.08789) + 2(0.26367) + 1(0.39551) + 0(0.23731) = 1.25$—on average, we would expect a student to guess about one and one quarter questions correctly.

For a Sample Space in Which Probabilities Are Assigned Empirically

The first step is to define an appropriate probability distribution, which is just a matter of matching up the probabilities to the corresponding values of the variable. After that, you can find the expected value by just summing the possible values, weighted by their probabilities: $E(X) = \sum_{i=1}^{N} x_i P(x_i)$.

For example, suppose that you find data on pet ownership in a particular neighborhood claiming that 22% of the neighborhood's households own no pets, 30% own one pet, 19% own two, 13% own three pets, 9% own four pets, 4% own five pets, 3% own six pets, and 1% own seven pets. The expected value for the number of pets in a household in the neighborhood is then (0)(0.22)+(1)(0.30)+(2)(0.19)+(3)(0.13)+(4)(0.09)+(5)(0.04)+(6)(0.03)+(7)(0.01)=1.88. The mean number of pets owned by a household in the neighborhood is 1.88; in 100 randomly selected households, you would expect to find a total of (100)(1.88) = 188 pets.

Expected Payoff for a Game of Chance

The expected payoff for a game of chance is simply the expected value for a participant's winnings. This can be calculated the same way the expected value is calculated of any probability distribution: by summing all the possible values for the winnings together, weighted by their probability. For example, suppose a participant in a particular lottery has a one in ten million chance of winning a million dollars—and otherwise wins nothing. Then the expected payoff is equal to $\left(\frac{1}{10{,}000{,}000}\right)(\$1{,}000{,}000) + \left(\frac{9{,}999{,}999}{10{,}000{,}000}\right)(\$0) = \$0.10$.

For a more complicated example, suppose a particular game of chance has a 1 in 10 chance of paying off \$3, a 1 in 20 chance of paying \$5, and a 1 in 100 chance of paying \$20. Then the expected payoff is $\left(\frac{1}{10}\right)(\$3) + \left(\frac{1}{20}\right)(\$5) + \left(\frac{1}{100}\right)(\$20) = \$0.75$. (We omit this time writing down the term corresponding to a payoff of \$0, since the probability of this term doesn't affect the result anyway.)

Making a Profit in a Game of Chance

In order to make a profit on a sweepstake, raffle, lottery, or other game of chance, it's necessary for the charge for participation to be more than the expected payoff—otherwise, you're likely to pay out as much or more to winners than you bring in. Of course, the more the charge for tickets, the higher the profits, but setting the cost too high is likely to discourage people from entering.

For example, suppose that a particular raffle gives out one prize worth \$5000, five prizes worth \$1000, and twenty prizes worth \$100. If the organizers of the raffle expect to sell five thousand raffle tickets, then the expected payoff is equal to $\left(\frac{1}{5000}\right)(\$5000) + \left(\frac{5}{5000}\right)(\$1000) + \left(\frac{20}{5000}\right)(\$100) = \$2.40$In order to make a profit, then, the organizers of the raffle have to charge more than \$2.40 for raffle tickets.

High Chance of a Low Payoff vs. Low Chance of a High Payoff

It is not always the case that a high chance of a low payoff is better than a low chance of a high payoff, or vice versa. To use probabilistic techniques to choose between a high chance of a low payoff and a low chance of a high payoff, it's necessary to compare the expected payoffs from each option.

For instance, suppose you have a choice between a 10% chance of getting \$100, or a 1% chance of getting \$2,000. The expected payoff from the first option is $(0.1)(\$100) + (0.9)(\$0) = \$10$, while the expected payoff from the second option is $(0.01)(\$2000) + (0.99)(\$0) = \$20$. In this case, the higher-risk option is the better choice. On the other hand, if you have to choose between a 25% chance of getting \$100 or a 1% chance of getting \$2000, then the expected payoff from the first option is $(0.25)(\$100) + (0.75)(\$0) = \$25$, and the expected payoff from the second option is $(0.01)(\$2000) + (0.99)(\$0) = \$20$. In this case, the better choice is the lower-risk option.

Using Probabilities to Choose Between Insurance Policies

To use probabilities to choose between insurance policies, you can calculate the expected net cost of each policy, depending on the estimated probability that the policy will be needed. Suppose you have a choice between two car insurance policies: one costs \$50 a month and pays \$1,000 in the case of a severe accident, while the other costs \$100 a month but pays \$4,000. If you estimate your chances in a given month of getting in a severe accident as 1 in 100, then the expected payoff for the first policy is $\left(\frac{1}{100}\right)(\$1000) = \$10$, minus the \$50 a month for a total monthly net cost of \$40. Similar calculations show that the second policy has an expected net monthly cost of \$60. The first policy, then, would be the better choice. However, if your chances of getting in an accident in a given month are 1 in 50, then the monthly cost of the first policy becomes \$30, and of the second \$20. Now the second policy is the better choice.

Using Probabilities to Make a Fair Decision Between a Number of Options

A random decision may be considered "fair" if all outcomes are equally likely. To make a fair decision, therefore, it is necessary to choose a procedure that will make all outcomes equally likely. Analyzing a model using the laws of probability can help determine whether or not that model really gives rise to equally probable outcomes.

For example, suppose it is necessary to randomly choose one of eleven people to receive some prize. One way that may come to mind is to roll two dice and sum the results. This yields a number between 2 and 12—eleven possible outcomes. However, these outcomes are not equally likely. The probability of rolling a 2 is only $\frac{1}{36}$, for example, but the probability of rolling a 7 is $\frac{1}{6}$—six times as likely! A procedure better suited to a fair decision would be drawing lots. For instance, eleven identical slips of paper could be labeled with the name of one of the possible outcomes, and one of the slips randomly drawn.

TRUE RANDOM NUMBER GENERATOR

A true random number generator should produce all possible outcomes with equal probability. Therefore, one can test the random number generator by performing a large number of trials and examining the distribution of the results. If indeed all possible results do seem to appear with equal frequency, then the random number generator may be truly random. If some results appear with significantly greater frequency than others, then the random generator is faulty. Technically, just the fact that the possible outcomes are equally likely does not necessarily make the random number generator random, however. Technically, for instance, it's possible for a random generator to produce all possible outcomes in a fixed order, cycling through them all before returning to the beginning of the sequence. Such a number generator would produce all outcomes with equal probability, but would not be random. Such cases may be more difficult to test for, but extreme examples like this one can often be easily spotted.

EXAMPLE USES OF PROBABILITY CONCEPTS

MEDICAL TESTING

Medical testing can be conducted by controlled experimental trials. Patients can be separated into several groups, some of which receive varying amounts of the medication or other treatment that is being tested, while the others receive a placebo (a substance with no effect so that the patients themselves do not know which are receiving the real treatment). Variables related to the effect of the treatment can then be observed and compared between the groups. Based on what we know about probability, there are some considerations that must be taken into account when conducting such a test. For one, the patients must be chosen and the groups assigned randomly, to avoid bias. If the groups are not initially similar to each other, then the difference in the results will not be meaningful. Also, before coming to conclusions based on the result, it is important to make sure that the differences between the groups are really significant, and that we do not assume that correlation proves causation.

PRODUCT MARKET RESEARCH

Market research is the practice of finding out what the public thinks about certain current or proposed products and services. This is usually done by means of a sample survey: it's impractical to survey every individual potential customer, so researchers choose a representative sample of possible customers and survey them as to their opinions on the product or service in question. Based on what we know about probability, there are some considerations that must be taken into account when conducting such research. For one, the survey participants must be chosen randomly, to avoid bias. If the survey participants are all from a similar subset of the population, they may not be a representative sample of the whole population of potential customers, and the results from the survey may not extend to the population as a whole. Also, before coming to conclusions based on the result, it is important to make sure that the differences between the responses to different products are really significant.

CHANGING THE GOALIE DURING A LOSING HOCKEY GAME

If there seem to be a lot of goals scored on the goalie on a hockey team, it may be tempting to replace to goalie to improve the circumstance. However, this may not always be the best choice, and probability considerations may allow for more consideration of the matter. One thing to consider is that correlation does not prove causation: the fact that there are more goals being scored against this goalie doesn't mean it's the goalie's fault. Perhaps there's a third variable in action; could the other defensive players have been put in the same time as the goalie, and are they allowing more goals to reach him? It's also important to consider whether the apparent increase in the number of goals being scored is really significant, or whether it might just be the result of random fluctuation. Of course, similar principles apply not only to the case of switching out goalies in hockey games; these same considerations apply to a wide variety of real-life situations.

Practice Test #1

Practice Questions

1. Which of the following boxplots correctly represents the data set shown below?

7, 8, 8, 11, 13, 14, 15, 15, 16, 17, 21, 23, 23, 24, 31, 35, 35, 35, 38, 38, 40, 45, 49, 50, 50

a.

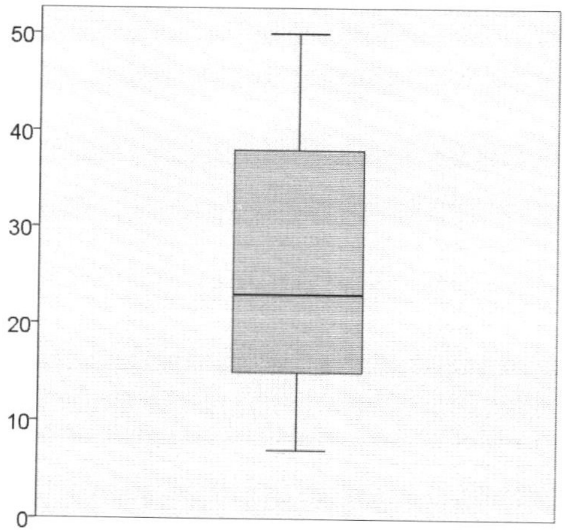

c.

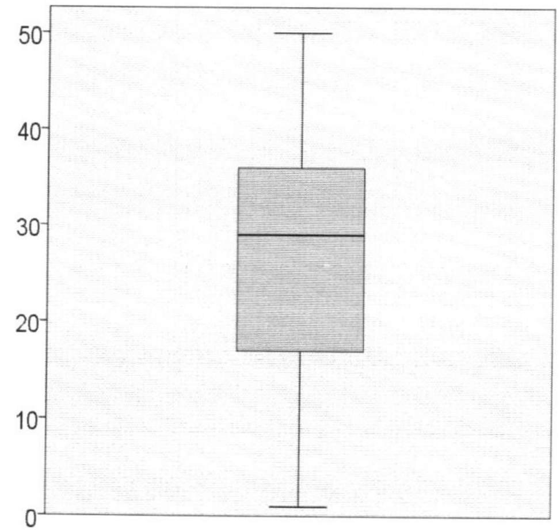

b.

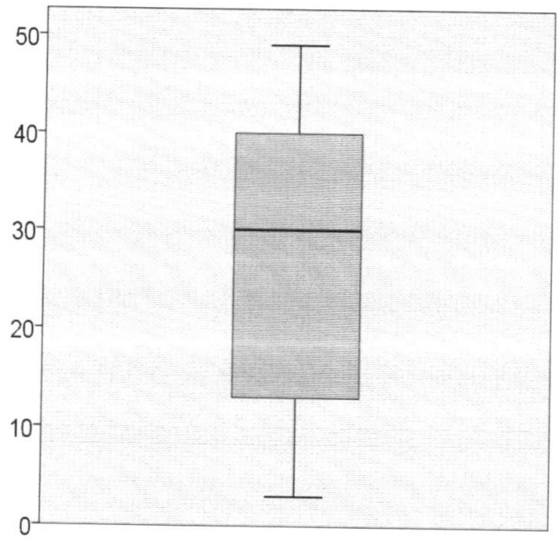

d.

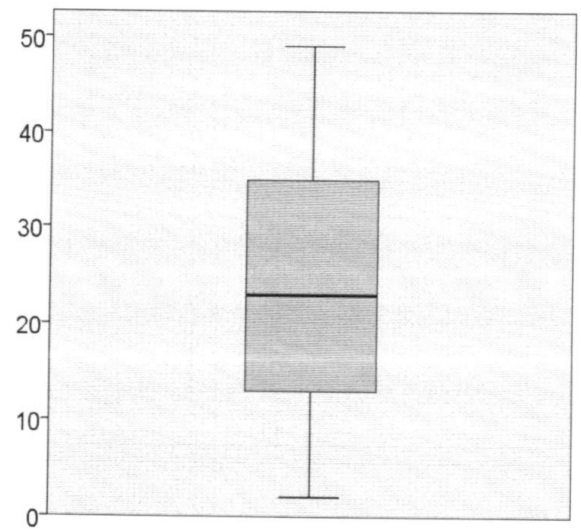

2. Given the parallel boxplots shown below, which of the following statements is true?

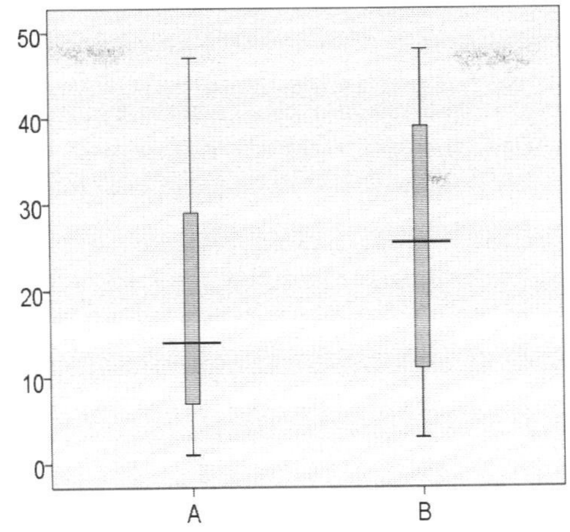

a. The boxplot for data set A has a larger interquartile range and a larger median.
b. The boxplot for data set B has a larger interquartile range and a larger median.
c. The boxplot for data set A has a larger interquartile range and a smaller median.
d. The boxplot for data set B has a larger interquartile range and a smaller median.

3. Given the double bar graph shown below, which of the following statements is true?

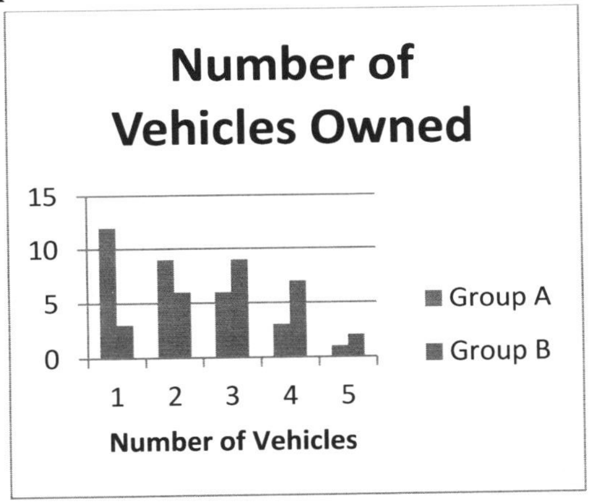

a. Group A is negatively skewed, while Group B is approximately normal.
b. Group A is positively skewed, while Group B is approximately normal.
c. Group A is approximately normal, while Group B is negatively skewed.
d. Group A is approximately normal, while Group B is positively skewed.

4. Which of the following data sets is represented by the histogram shown below?

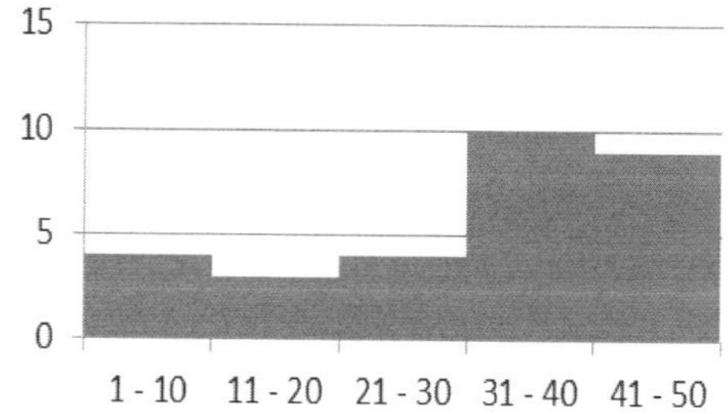

a. 6,7, 7, 8, 12, 13, 18, 22, 26, 27, 28, 31, 36, 36, 36,
36, 37, 38, 38, 38, 39, 41, 41, 41, 42, 42, 43, 43, 48, 49
b. 2, 3, 4, 4, 8, 9, 10, 10, 11, 15, 19, 21, 21, 21, 21,
22, 24, 24, 25, 29, 31, 33, 40, 43, 45, 46, 48, 48, 49, 50
c. 1, 2, 8, 8, 9, 9, 12, 13, 15, 15, 16, 17, 18, 19, 19,
21, 29, 31, 31, 35, 36, 38, 41, 42, 42, 45, 46, 47, 47, 49
d. 1, 4, 4, 5, 10, 10, 10, 12, 12, 16, 20, 22, 23, 23, 25,
27, 31, 31, 37, 39, 40, 41, 41, 43, 45, 46, 48, 49, 49, 50

5. The statistics scores for Sections 101 and 102 are shown below. Which of the following statements accurately compare the center and variability of the sets of scores?

Section 101	60	62	63	66	70	82	83	85	88	90	90	91	95	95	97
Section 102	65	65	67	69	69	70	73	76	77	79	86	90	95	98	98

a. The statistics scores for Section 101 show a greater center and variability.
b. The statistics scores for Section 101 show a greater center but smaller variability.
c. The statistics scores for Section 101 show a smaller center but greater variability.
d. The statistics scores for Section 101 show a smaller center and variability.

6. Given the histograms shown below, which of the following statements is true?

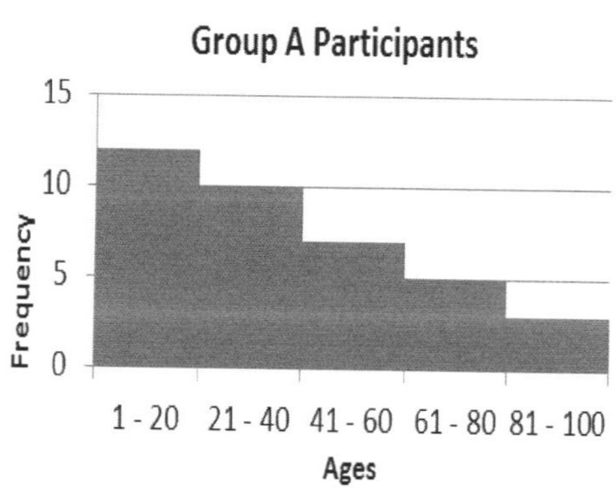

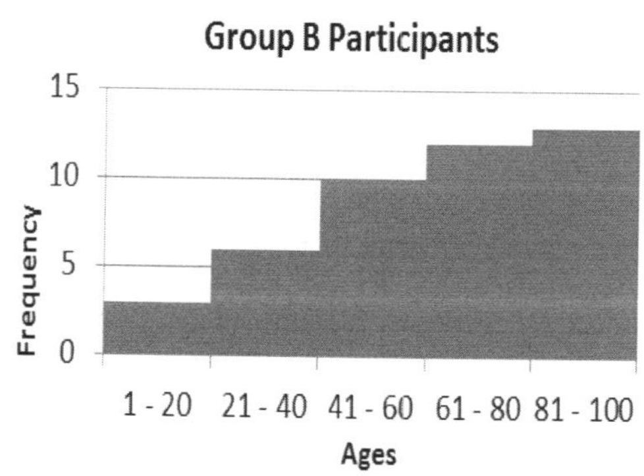

a. Group A is negatively skewed and has a mean that is less than the mean of Group B.
b. Group A is positively skewed and has a mean that is more than the mean of Group B.
c. Group B is negatively skewed and has a mean that is more than the mean of Group A.
d. Group B is positively skewed and has a mean that is less than the mean of Group A.

7. Given a normal distribution, which of the following best represents the area under the curve to the right of a z-score of 1.38?

a. 0.03
b. 0.08
c. 0.12
d. 0.15

8. Given μ = 18 and σ = 2, which of the following best represents the proportion of scores falling above 19?

a. 31%
b. 18%
c. 41%
d. 22%

9. Which of the following best represents the area under the curve of a normal distribution, falling between the z-scores of −0.8 and 1.2?

a. 62%
b. 67%
c. 72%
d. 77%

10. Given the table below, which of the following best represents the probability that a student is enrolled at TAMU or prefers lattes?

	Latte	Cappuccino	Frappuccino	Total
TAMU	350	225	175	750
NMSU	325	300	275	900
Total	675	525	450	1,650

a. 55%
b. 60%
c. 65%
d. 70%

11. The high school GPAs and SAT scores, for a random group of students, is shown in the table below. According to the least-squares regression line, which of the following best represents the SAT score that may be predicted by a high school GPA of 93?

GPA	70	77	79	82	85	86	87	88	89	90	92	99
SAT	800	1300	700	1400	2100	2100	2250	1800	1950	1850	1500	2000

a. 1850
b. 1900
c. 1950
d. 2000

12. Which of the following correlation coefficients represents the weakest correlation?

a. 0.3
b. −0.1
c. 0.4
d. −0.9

13. Which of the following best represents the line of best fit for the data shown in the scatter plot below?

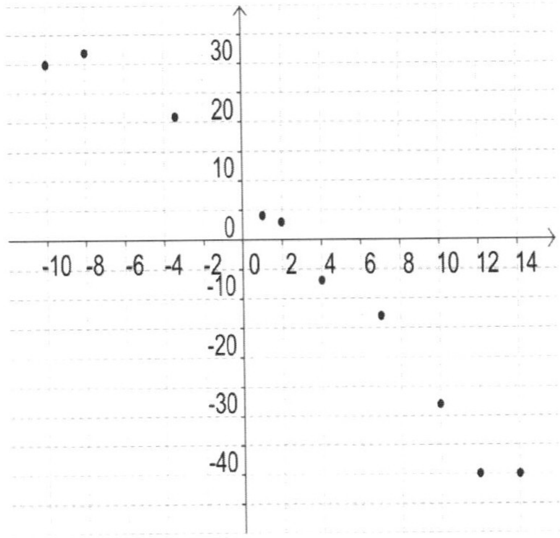

a. $y = -3.2x + 5.1$
b. $y = -2.4x + 10.3$
c. $y = -4.7x - 5.3$
d. $y = -8.2x + 3.4$

14. Aaron's cumulative coffee expenses over the course of *x* years are represented by the scatter plot below.

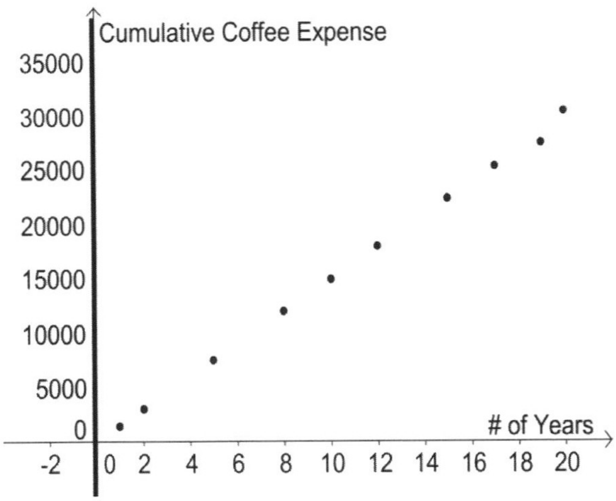

Which of the following best represents the increase in cumulative coffee expenses each year?

a. $1,478
b. $1,098
c. $1,892
d. $928

15. Given the data shown in the table below, which of the following best represents the correlation coefficient, *r*?

x	1	2	2	2	3	3	3	3	4	8	10	10	10	10	10
y	11	21	26	26	12	12	22	24	26	6	2	2	11	20	24

a. −0.48
b. −0.77
c. 0.86
d. 0.32

16. Suppose two variables show a correlation of 0.9. Which of the following statements *best* describes the relationship?

a. The variables show a strong correlation.
b. The variables show some correlation, which is neither weak nor strong.
c. The variables show a weak correlation.
d. The variables show a perfect correlation.

17. A student scores an 82 on a mathematics exam. The class average is 80 with a standard deviation of 4 points. Which of the following statements is true?

a. With a p-value greater than 0.05, the student's score is significantly higher than the class average.
b. With a p-value less than 0.05, the student's score is significantly higher than the class average.
c. With a p-value greater than 0.05, the student's score is not significantly higher than the class average.
d. With a p-value less than 0.05, the student's score is not significantly higher than the class average.

18. A dairy company claims to include 8 ounces of milk in each container. In a random sample of 50 containers, the mean number of ounces was 7.6 with a standard deviation of 0.2 ounces. Which of the following statements is true?

a. With the absolute value of the t-value less than the critical t-value, the mean number of ounces found in the random sample of containers is significantly lower.
b. With the absolute value of the t-value more than the critical t-value, the mean number of ounces found in the random sample of containers is significantly lower.
c. With the absolute value of the t-value less than the critical t-value, the mean number of ounces found in the random sample of containers is not significantly lower.
d. With the absolute value of the t-value more than the critical t-value, the mean number of ounces found in the random sample of containers is not significantly lower.

19. Elizabeth rolls a standard die, labeled 1 to 6, 100 times. Which of the following experimental outcomes is *not* likely?

a. 67 rolls will show a number greater than 2.
b. 50 rolls will show an even number.
c. 75 rolls will show a number less than 4.
d. 33 rolls will show a number less than 3.

20. A professor reports a mean average in his Educational Psychology course of 76 with a standard deviation of 5 points. His most recent class, with a total of 24 enrolled students, shows an average of 80. Which of the following statements is true?

a. With a *p*-value greater than 0.05, the average for his most recent class is significantly higher than the reported mean average.
b. With a p-value less than 0.05, the average for his most recent class is significantly higher than the reported mean average.
c. With a p-value greater than 0.05, the average for his most recent class is not significantly lower than the reported mean average.
d. With a p-value less than 0.05, the average for his most recent class is not significantly lower than the reported mean average.

21. A dog food manufacturing company claims to include 5 pounds of dog food in each bag with a standard deviation of 0.3 pounds. Suppose a randomly selected bag contains only 4.7 pounds of dog food. Which of the following statements is true?

a. With a *p*-value less than 0.05, the amount of dog food in the randomly selected bag is significantly different from the amount of dog food the company claims to include in each bag.
b. With a p-value greater than 0.05, the amount of dog food in the randomly selected bag is significantly different from the amount of dog food the company claims to include in each bag.
c. With a p-value less than 0.05, the amount of dog food in the randomly selected bag is not significantly different from the amount of dog food the company claims to include in each bag.
d. With a p-value greater than 0.05, the amount of dog food in the randomly selected bag is not significantly different from the amount of dog food the company claims to include in each bag.

22. Which of the following describes an experiment?

a. The final math grades for a group of students passing through each year of elementary school are examined.
b. The health trends of smokers in a small random sample are examined.
c. Citizens in a local community are surveyed to determine concerns related to the next election.
d. The effects of a new drug are tested on a group of participants.

23. When determining who will be included on a survey mailing list, a researcher chooses every 10th name in a phone book. Which random sampling technique best describes the method the researcher is using?

a. Stratified random sampling
b. Cluster sampling
c. Simple random sampling
d. Systematic random sampling

24. A survey is given to 50 different groups of random participants within a population. The means are plotted and reveal an approximate sampling distribution mean of 65. Which of the following most likely represents the population mean?

a. 35
b. 50
c. 65
d. 80

25. Suppose a sample of 50 participants shows a mean of 100 and a standard deviation of 5. With a 95% level of confidence, which of the following best represents the margin of error?

a. 0.8
b. 1.4
c. 1.9
d. 2.5

26. A sample of 100 participants reveals a mean of 120 and a standard deviation of 10. Which confidence interval below contains the actual population mean in 95% of sample cases?

a. (118.04, 121.96)
b. (106.08, 109.32)
c. (120.96, 123.48)
d. (112.12, 115.86)

27. An instructor uses two different instructional approaches with his two classes. He wishes to compare the effectiveness of the two approaches. The first class has 35 students. The class average of the final grades is 80 with a standard deviation of 5. The other class has 40 students and reports a class average of final grades of 82 with a standard deviation of 3. If the instructor uses a 0.05 level of significance, which of the following statements is true?

a. The difference between the class averages is significant as evidenced by a z-score of approximately −2.1.
b. The difference between the class averages is significant as evidenced by a z-score of approximately 1.6.
c. The difference between the class averages is not significant as evidenced by a z-score of approximately −2.1.
d. The difference between the class averages is not significant as evidenced by a z-score of approximately 1.6.

28. What is the probability that Ana draws a jack, replaces the card, and then draws an ace card?

a. $\frac{1}{52}$
b. $\frac{3}{208}$
c. $\frac{1}{169}$
d. $\frac{17}{52}$

29. Edward draws a card from a deck of cards, does not replace it, and then draws another card. What is the probability that he draws a heart and then a spade?

a. $\frac{1}{16}$
b. $\frac{1}{2}$
c. $\frac{1}{17}$
d. $\frac{13}{204}$

30. Nicolle spins a spinner with eight equally-spaced sections labeled 1 - 8. What is the probability that the spinner lands on a number greater than 5 or on 1?

a. $\frac{1}{2}$
b. $\frac{3}{16}$
c. $\frac{3}{64}$
d. $\frac{5}{8}$

31. Andrew draws a marble out of a bag, containing 2 red marbles, 3 blue marbles, and 5 green marbles. What is the probability that he draws either a red or a green marble?

a. $\frac{1}{10}$
b. $\frac{2}{3}$
c. $\frac{1}{4}$
d. $\frac{7}{10}$

32. Monique rolls a die. What is the probability that she rolls a 3 or a number greater than 2?

a. $\frac{5}{6}$
b. $\frac{1}{9}$
c. $\frac{2}{3}$
d. $\frac{2}{9}$

33. A bag contains 8 red marbles, 3 blue marbles, and 4 green marbles. What is the probability Carlos draws a red marble, does not replace it, and then draws another red marble?

a. $\frac{4}{15}$
b. $\frac{32}{105}$
c. $\frac{64}{225}$
d. $\frac{2}{15}$

34. The probability that Bryan goes to the store is 0.18. The probability that he goes to the movies is 0.34. The probability that he goes to the store, given that he goes to the movies, is 0.46. What is the probability he goes to the store or to the movies?

a. 32.38%
b. 36.36%
c. 40.28%
d. 44.46%

35. When considering the probabilities of independent events, A and B, which of the following is true?

I. $P(A) = P(A|B)$
II. $P(B) = P(A|B)$
III. $P(A) = P(B|A)$
IV. $P(B) = P(B|A)$

a. I only
b. I and IV only
c. II and III only
d. I, II, III, and IV

36. The probability that Alisha chooses a philosophy class is 0.30. The probability that she chooses a qualitative methods class, given that she chooses a philosophy class, is 0.40. Finally, the probability she chooses a philosophy class or a qualitative methods class is 0.62. What is the probability she chooses a qualitative methods class?

a. 0.32
b. 0.38
c. 0.44
d. 0.52

37. Given the two-way frequency table shown below, what is the probability a citizen is a Democrat or a Senior?

	Young Adult	**Middle-Aged**	**Senior**	**Total**
Democrat	52	82	60	194
Republican	36	102	72	210
Total	88	184	132	404

a. $\frac{133}{202}$
b. $\frac{135}{204}$
c. $\frac{121}{202}$
d. $\frac{123}{200}$

38. Suppose Eric draws three marbles from a bag containing 2 yellow marbles, 3 green marbles, and 5 blue marbles. Each time he draws a marble, he checks the color and then puts it back in the bag. Which of the following best represents the probability that he does not draw a green marble?

a. 23%
b. 27%
c. 29%
d. 34%

39. Jonathan rolls a die two times. What is the probability that he gets exactly two 4's?

a. $\frac{1}{18}$
b. $\frac{1}{24}$
c. $\frac{1}{30}$
d. $\frac{1}{36}$

40. Which of the following represents the expected value of the number of tails Adam will get after tossing a coin 6 times?

a. 2
b. 3
c. 6
d. 12

41. What is the expected value of drawing a card from a deck when the cards are labeled 1 – 5?

a. 1.5
b. 2
c. 2.5
d. 3

42. Suppose Ashley will receive a $6,000 scholarship if she chooses University A, a $4,500 scholarship if she chooses University B, and a $5,500 scholarship if she chooses University C. The probabilities that she will attend each university are equal. Which of the following best represents the expected value for the scholarship amount she will receive?

a. $4,833
b. $5,155
c. $5,333
d. $5,525

43. A politician will receive funds from donors if he agrees to support a particular cause. The politician will receive $100,000 from Donor A, $150,000 from Donor B, and $180,000 from Donor C. There is a 0.32 probability he will support Donor A's cause, a 0.40 probability he will support Donor B's cause, and a 0.28 probability he will support Donor C's cause. What is the average amount of funds he can expect to receive?

a. $128,200
b. $132,800
c. $138,600
d. $142,400

44. If Henry rolls a die 8 times, what is the expected sum?

a. 22
b. 24
c. 26
d. 28

45. Nia spends $5 on a raffle ticket. If she wins, she will win $500. A total of 250 raffle tickets were sold. What is the expected value?

a. −$2
b. −$3
c. −$5
d. −$7

46. Hannah draws two cards from a deck with cards labeled 1 – 6. What is the expected number of 3's?

a. $\frac{1}{3}$
b. $\frac{1}{4}$
c. $\frac{1}{6}$
d. $\frac{1}{8}$

47. Given the probability distribution table shown below, what is the expected value?

x	f(x)
0	0.23
1	0.21
2	0.17
3	0.14
4	0.12
5	0.09
6	0.04

a. 1.82
b. 1.96
c. 2.02
d. 2.14

48. Bryan draws 3 cards from a deck of cards. Which of the following best represents the expected value of drawing an ace card?

a. 0.15
b. 0.19
c. 0.23
d. 0.27

49. Robin draws 2 cards from a deck of cards. If the probability distribution of drawing a jack were to be graphed using a histogram, which of the following *y*-values would correspond to the *x*-value draws of a spade, namely 0, 1, and 2?

a. 0.56, 0.38, 0.06
b. 0.32, 0.44, 0.24
c. 0.48, 0.30, 0.22
d. 0.42, 0.38, 0.20

50. Donald rolls a die. If the die lands on a 1 or 2, he wins $5.00. If the die lands on a 3, he loses $1.00. If the die lands on a 4, 5 or 6, he loses $2.50. Which of the following statements is true?

I. He can expect to win $0.25 after 1 roll.
II. He can expect to lose $0.50 after 1 roll.
III. His expected winnings for 50 rolls sum to $12.50.
IV. His expected losses for 50 rolls sum to $25.00.

a. I only
b. I and III only
c. II and IV only
d. IV only

Answers and Explanations

Math.Content.S.ID.1

1. A: The minimum and maximum values are 7 and 50, respectively. The median is 23, while the first and third quartiles are 15 and 38. The boxplot for Choice A correctly represents these five values.

Math.Content.S.ID.2

2. B: The boxplot for data set B shows a larger range between the first and third quartiles. In addition, its median is approximately 25, which is higher than the median of data set A.

Math.Content.S.ID.3

3. B: Data is said to be positively skewed when there are a higher number of lower values, indicating data that is skewed right. An approximately normal distribution shows an increase in frequency, followed by a decrease in frequency, of approximately the same rate.

Math.Content.S.ID.1

4. A: The data set, shown in Choice A, correctly represents a frequency of 4 for the interval 1 – 10, a frequency of 3 for 11 – 20, a frequency of 4 for 21 – 30, a frequency of 10 for 31 – 40, and a frequency of 9 for 41 – 50.

Math.Content.S.ID.3

5. A: The scores for Section 101 have a mean of approximately 81.1, with a standard deviation of approximately 13.2. The scores for Section 102 have a mean of approximately 78.5, with a standard deviation of approximately 12.0. Thus, the center and variability of scores for Section 101 are greater.

Math.Content.S.ID.2

6. C: Group B is negatively skewed since there are more high scores. With more high scores, the mean for Group B will be higher.

Math.Content.S.ID.4

7. B: Using a standard z-table, the area from the mean to z is shown to be 0.4162. This area should be subtracted from 0.5 in order to show the area to the right of a z-score of 1.38. Thus, the area to the right of the z-score is 0.0838, or approximately 0.08. A z-table, including the smaller portion of area under a curve may also be used. This area is reported as 0.0838.

Math.Content.S.ID.4

8. A: The z-score may be represented as $z = \frac{19-18}{2}$ or $z = \frac{1}{2}$. Using a standard z-table, the area from the mean to z is shown to be 0.1915. Subtraction of this area from 0.5 gives 0.3085. Thus, the proportion of scores falling above 19 is approximately 31%. A z-table, including the smaller portion of area under a curve may also be used. This area is reported as 0.3085.

Math.Content.S.ID.4

9. B: A standard z-table shows an area of 0.2881 between a z-score of -0.8 and the mean. It also shows an area of 0.3849 between the mean and a z-score of 1.2. The sum of the two areas is 0.673. Thus, the area between the two z-scores is approximately 67%. Using a table that reports the mean to z, the two mean-to-z areas may be extracted and summed. You may also add the smaller portions and subtract the sum from 1.

Math.Content.S.ID.5

10. C: To find the probability that a student is enrolled at TAMU or that a student prefers lattes, the addition rule needs to be used. The addition rule adds the probabilities of two independent events and subtracts the probability that both events are true to avoid double counting. The probability may be written as $P(A \text{ or } B) = \frac{750}{1{,}650} + \frac{675}{1{,}650} - \frac{350}{1{,}650}$, which simplifies to $P(A \text{ or } B) = \frac{1{,}075}{1{,}650} \approx 65\%$. Therefore, the probability is approximately 65%.

Math.Content.S.ID.6a

11. D: The least-squares regression line is written in the form, $y = ax + b$, where a represents the slope and b represents the y-intercept. Entering the scores as lists into a graphing calculator allows you to calculate the least-squares regression line. Choose Stat, Calc, and then LinReg, in order to find the values of a and b. Doing so gives an approximate least-squares regression line of $y = 47.68x - 2423$. Substituting a GPA of 93 for the x-value gives a predicted SAT score of approximately 2011. Thus, 2000 is the best choice for an SAT score, predicted by a GPA score of 93.

Math.Content.S.ID.6b

12. B: Weak correlation coefficients are those with absolute values close to 0. Since −0.1 has an absolute value of 0.1 and 0.1 is closer to 0 than any of the absolute values of the other correlation coefficients, it is the weakest.

Math.Content.S.ID.6c

13. A: The correlation is negative with a slope of approximately −3 and a y-intercept of approximately 5. Choices B and C can be eliminated, since a y-intercept of 10.3 is too high and a y-intercept of −5.3 is too low. Choice D may also be eliminated since the slope of −8.2 is too steep for the slope shown on the graph.

Math.Content.S.ID.7

14. A: The slope represents the increase in coffee expenses after each year. The slope can be calculated by choosing any two points. For this explanation, we will use the points, (17, 25000) and (5, 7500). The slope may be written as $\frac{7500-25{,}000}{5-17}$, which is approximately 1,458. Thus, $1,478 is the best approximation for the increase in coffee expenses each year.

Math.Content.S.ID.8

15. A: The data may be entered into lists, using the graphing calculator. Next, choose Stat, Calc, and then LinReg. The resulting r-value will be displayed. In this case, the r-value is approximately −0.48.

Math.Content.S.ID.9

16. A: A correlation of 0.9 is a strong correlation. Recall the closer a correlation coefficient is to −1 or 1, the stronger the correlation. A correlation of −1 or 1 is perfect.

Math.Content.S.IC.1

17. C: The z-score may be calculated, using the formula, $z = \frac{X-\mu}{\sigma}$. Substituting a value of 82 for X, 80 for μ, and 4 for σ gives $z = \frac{82-80}{4}$ or $z = \frac{1}{2}$. A z-score of 0.5 has a p-value of approximately 0.31, which is larger than 0.05. Therefore, there is not a statistically significant difference between the score and the class average.

Math.Content.S.IC.1

18. B: Since only the sample standard deviation is known, a t-distribution will be used. When comparing a sample mean to a population mean, use the formula, $t = \frac{\bar{X}-\mu}{\frac{s}{\sqrt{n}}}$, where $\bar{X}$ represents the sample mean, μ represents the population mean, s represents the sample standard deviation, and n represents the sample size. Substituting 7.6 for $\bar{X}$, 8 for μ, 0.2 for s, and 50 for n gives $t = \frac{7.6-8}{\frac{0.2}{\sqrt{50}}}$, or $t = -14.14$. The absolute value of −14.14, namely 14.14, is more than the critical t-value of approximately 1.676. Note that the critical t-value for a one-tailed test with a 0.05 level of significance for 50 degrees of freedom was used. The actual degrees of freedom is 49, but 50 was used because it is the closest value in the table. Since the absolute value of the actual t-value is more than the critical t-value, the mean of the random sample may be interpreted as significantly lower.

Math.Content.S.IC.2

19. C: The theoretical probability of rolling a number less than 4 is the same as the theoretical probability of rolling an even number; the probability is $\frac{1}{2}$. The expected value is equal to the product of the number of rolls of 100 and the probability of $\frac{1}{2}$. Thus, it is likely that 50 of the rolls will show a number less than 4, not 75 of the rolls.

Math.Content.S.IC.1

20. B: A z-distribution will be used since the population standard deviation is known. When comparing a sample mean to a population mean, use the formula, $z = \frac{\bar{X}-\mu}{\frac{\sigma}{\sqrt{n}}}$, where $\bar{X}$ represents the sample mean, μ represents the population mean, σ represents the population standard deviation, and n represents the sample size. Substituting 80 for $\bar{X}$, 76 for μ, 5 for σ, and 24 for n gives $z = \frac{80-76}{\frac{5}{\sqrt{24}}}$, or $z \approx 3.9$. The p-value for a z-score of 3.9 is less than 0.001. Therefore, there is a significant difference between the means as evidenced by a p-value less than 0.05.

Math.Content.S.IC.1

21. D: A z-distribution will be used since the population standard deviation is known. The formula, $z = \frac{X-\mu}{\sigma}$, will be used. Substituting 4.7 for X, 5 for μ, and 0.3 for σ gives: $z = \frac{4.7-5}{0.3}$, where $z = -1$. The probability for a z-score with an absolute value of 1 is approximately 0.16. Since the p-value is greater than 0.05, there is not a significant difference between the amount of dog food in the randomly selected bag and the mean amount per bag claimed by the company.

Math.Content.S.IC.3

22. D: Only Choice D introduces an intervention to be studied. Thus, Choice D represents an experiment.

Math.Content.S.IC.3

23. D: When choosing every nth name on a list or roster, systematic random sampling is used.

Math.Content.S.IC.4

24. C: The mean of the sampling distribution approaches (and approximates) the actual population mean. Therefore, the population mean will be close to 65.

Math.Content.S.IC.6

25. B: The formula for finding the margin of error is $E = z_{\alpha/2} \cdot \frac{\sigma}{\sqrt{n}}$, where z represents the z-score for a 95% degree of confidence for a two-tailed test, σ represents the standard deviation, and n

represents the sample size. When the sample size is larger than 30, the sample standard deviation may be substituted for the population standard deviation σ. Thus, the margin of error may be represented as $E = 1.96 \cdot \frac{5}{\sqrt{50}}$, which is approximately 1.4.

Math.Content.S.IC.4

26. A: The margin of error may be represented as $E = 1.96 \cdot \frac{10}{\sqrt{100}}$, or 1.96. Subtracting 1.96 from 120 gives 118.04. Adding 1.96 to 120 gives 121.96. Thus, the population mean will likely fall within the 95% confidence interval of (118.04, 121.96).

Math.Content.S.IC.5

27. A: The problem asks you to compare two means. A two-sample z-test may be used. The given means, standard deviations, and sample sizes may be entered into a graphing calculator, in order to determine the critical z-value and p-value. Choose Stat, Tests, and 2-SampZTest. The resulting z-score is approximately −2.1 with a p-value of approximately 0.04. Therefore, there is a significant difference between the class averages.

Math.Content.S.CP.2

28. C: The probability of independent events, A and B, may be found using the formula, $P(A \text{ and } B) = P(A) \cdot P(B)$. Thus, the probability she draws a jack, replaces it, and then draws an ace card may be represented as $P(A \text{ and } B) = \frac{4}{52} \cdot \frac{4}{52}$, which simplifies to $P(A \text{ and } B) = \frac{1}{169}$.

Math.Content.S.CP.1

29. D: Since he does not replace the first card, the events are dependent. The sample space will decrease by 1 for the second draw because there will be one less card to choose from. Thus, the probability may be written as $P(A \text{ and } B) = \frac{13}{52} \times \frac{13}{51}$, or $P(A \text{ and } B) = \frac{169}{2{,}652}$, which simplifies to $\frac{13}{204}$.

Math.Content.S.CP.7

30. A: The probability of mutually exclusive events, A or B, occurring may be written as $P(A \text{ or } B) = P(A) + P(B)$. Thus, $P(A \text{ or } B) = \frac{3}{8} + \frac{1}{8}$ or $\frac{1}{2}$.

Math.Content.S.CP.7

31. D: The probability of mutually exclusive events, A or B, occurring may be written as $P(A \text{ or } B) = P(A) + P(B)$. Thus, $P(A \text{ or } B) = \frac{2}{10} + \frac{5}{10}$ or $\frac{7}{10}$.

Math.Content.S.CP.6

32. C: The probability of non-mutually exclusive events, A or B, occurring may be written as $P(A \text{ or } B) = P(A) + P(B) - P(A \text{ and } B)$. Thus, $P(A \text{ or } B) = \frac{1}{6} + \frac{4}{6} - \frac{1}{6}$ or $\frac{2}{3}$.

Math.Content.S.CP.5

33. A: The events are dependent, since the first marble was not replaced. The sample space of the second draw will decrease by 1 because there will be one less marble to choose. The number of possible red marbles for the second draw will also decrease by 1. Thus, the probability may be written as $P(A \text{ and } B) = \frac{8}{15} \cdot \frac{7}{14}$. The probability he draws a red marble, does not replace it, and draws another red marble is $\frac{4}{15}$.

Math.Content.S.CP.8

34. B: The probability of events, A or B, may be written as $P(A \text{ or } B) = P(A) + P(B) - P(A \text{ and } B)$. The probabilities of A and B are given, as is the probability of A, given B, so first use the formula $P(A \text{ and } B) = P(B) \cdot P(A|B)$. Thus, $P(A \text{ and } B) = 0.34 \cdot 0.46 = 0.1564$. Substituting all probabilities gives 0.18 + 0.34 – 0.1564, which equals 0.3636. Thus, the probability that he goes to the store or to the movies is 36.36%.

Math.Content.S.CP.3

35. B: When events, A and B, are independent, the probability of A is equal to the probability of A given B. In addition, the probability of B is equal to the probability of B given A.

Math.Content.S.CP.8

36. C: The problem may be solved by writing $P(phil \text{ or } qual) = P(phil) + P(qual) - P(phil \text{ and } qual)$. The probability for philosophy or qualitative and the probability for philosophy may be substituted giving $0.62 = 0.30 + P(qual) - P(phil \text{ and } qual)$. Next find the probability that she chooses both philosophy and qualitative by using the formula, $P(phil \text{ and } qual) = P(phil) \cdot P(qual|phil)$. Substituting the probabilities of 0.30 and 0.40 gives $P(phil \text{ and } qual) = 0.30 \cdot 0.40$, or 0.12. Now, the probability of 0.12 may be substituted into the original equation, giving $0.62 = 0.30 + P(qual) - 0.12$. Solving for $P(qual)$ gives a probability of 0.44. Thus, there is a 0.44 probability that Alisha chooses a qualitative methods class.

Math.Content.S.CP.4

37. A: The probability of events, D or S, may be written as $P(D \text{ or } S) = P(D) + P(S) - P(D \text{ and } S)$. Thus, the probability a citizen is a Democrat or Senior may be represented as $P(D \text{ or } S) = \frac{194}{404} + \frac{132}{404} - \frac{60}{404}$ or $\frac{133}{202}$.

Math.Content.S.CP.9

38. D: The binompdf function of a graphing calculator may be used to find the probability of not drawing any green marbles. Choose 2nd Vars and then binompdf. Press Enter. List 3, $\frac{3}{10}$, and 0, within the open parenthesis. The first value represents the number of marbles drawn, the fraction represents the probability of drawing a green marble, and 0 represents the number of green marbles, seen as successes. Thus, the probability of not drawing a green marble is approximately 34%.

Math.Content.S.CP.9

39. D: The following may be entered into a graphing calculator: binompdf$\left(2, \frac{1}{6}, 2\right)$, which equals $\frac{1}{36}$.

Math.Content.S.MD.1

40. B: The number of tails he can expect after 6 coin tosses is equal to the product of the probability of getting tails on one coin toss and the number of coin tosses. Thus, the expected value is $\frac{1}{2} \cdot 6$, or 3.

Math.Content.S.MD.2

41. D: The expected value is equal to the sum of the products of each card value and its probability. Thus, the expected value is $\left(1 \cdot \frac{1}{5}\right) + \left(2 \cdot \frac{1}{5}\right) + \left(3 \cdot \frac{1}{5}\right) + \left(4 \cdot \frac{1}{5}\right) + \left(5 \cdot \frac{1}{5}\right)$, which equals 3.

Math.Content.S.MD.4

42. C: The expected value is equal to the sum of the products of the scholarship amounts and probability she will attend each college, or $\frac{1}{3}$. Thus, the expected value is $\left(6000 \cdot \frac{1}{3}\right) + \left(4500 \cdot \frac{1}{3}\right) + \left(5500 \cdot \frac{1}{3}\right)$, which equals 5,333. So, she can expect to receive \$5,333.

Math.Content.S.MD.4

43. D: The expected value is equal to the sum of the products of the amount the politician will receive from each donor and the probability the politician will support the donor's cause. Thus, the expected value is $(100{,}000 \cdot 0.32) + (150{,}000 \cdot 0.4) + (180{,}000 \cdot 0.28)$, which equals 142,400. So, the politician can expect to receive \$142,400.

Math.Content.S.MD.3

44. D: The expected value of one die roll is 3.5 $\left(\left(0 \cdot \frac{1}{6}\right) + \left(1 \cdot \frac{1}{6}\right) + \left(2 \cdot \frac{1}{6}\right) + \left(3 \cdot \frac{1}{6}\right) + \left(4 \cdot \frac{1}{6}\right) + \left(5 \cdot \frac{1}{6}\right) + \left(6 \cdot \frac{1}{6}\right)\right)$. Thus, the expected sum of 8 rolls will be equal to the product of the expected value of one roll and 8, or 28.

Math.Content.S.MD.5a

45. B: The expected value is equal to the sum of the product of the amount of the net profit, or \$495, and the probability of winning and the product of the amount of the loss, or −\$5, and the probability of losing. The expected value is $\left(495 \cdot \frac{1}{250}\right) + \left((-5) \cdot \frac{249}{250}\right)$, which equals −3. Thus, the expected value is −\$3.

Math.Content.S.MD.2

46. A: The table below may be used to calculate the expected value:

x	f(x)
0	$\frac{\binom{1}{0}\binom{5}{2}}{\binom{6}{2}}$
1	$\frac{\binom{1}{1}\binom{5}{1}}{\binom{6}{2}}$
2	$\frac{\binom{1}{2}\binom{5}{0}}{\binom{6}{2}}$

Thus, the expected value is $\left(0 \cdot \frac{10}{15}\right) + \left(1 \cdot \frac{5}{15}\right) + \left(2 \cdot \frac{0}{15}\right)$, which equals $\frac{1}{3}$.

Math.Content.S.MD.4

47. D: The expected value is equal to the sum of the products of the probabilities and their x-values. The expected value is $(0 \cdot 0.23) + (1 \cdot 0.21) + (2 \cdot 0.17) + (3 \cdot 0.14) + (4 \cdot 0.12) + (5 \cdot 0.09) + (6 \cdot 0.04)$, which equals 2.14.

Math.Content.S.MD.7

48. C: The expected value of drawing an ace card is approximately 0.23. The table below may be used to calculate the expected value:

X	F(x)
0	$\frac{\binom{4}{0}\binom{48}{3}}{\binom{52}{3}}$
1	$\frac{\binom{4}{1}\binom{48}{2}}{\binom{52}{3}}$
2	$\frac{\binom{4}{2}\binom{48}{1}}{\binom{52}{3}}$
3	$\frac{\binom{4}{3}\binom{48}{0}}{\binom{52}{3}}$

Thus, the expected value is $\left(0 \cdot \frac{17296}{22{,}100}\right) + \left(1 \cdot \frac{4512}{22{,}100}\right) + \left(2 \cdot \frac{288}{22{,}100}\right) + \left(3 \cdot \frac{4}{22{,}100}\right)$, which is approximately 0.23.

Math.Content.S.MD.1

49. A: The following table may be used to compute the probabilities:

X	F(x)
0	$\frac{\binom{13}{0}\binom{39}{2}}{\binom{52}{2}} = \frac{741}{1326}$
1	$\frac{\binom{13}{1}\binom{39}{1}}{\binom{52}{2}} = \frac{507}{1326}$
2	$\frac{\binom{13}{2}\binom{39}{0}}{\binom{52}{2}} = \frac{78}{1326}$

Thus, the corresponding y-values are approximately 0.56, 0.38, and 0.06.

Math.Content.S.MD.6

50. B: The expected value is equal to the sum of the product of the probabilities of the rolls and the amount he will win or lose. The probability of rolling a 1 or 2 is $\frac{2}{6}$. The probability of rolling a 3 is $\frac{1}{6}$. The probability of rolling any other number is $\frac{3}{6}$. Since he will win \$5.00 if he rolls a 1 or 2, lose \$1.00 if he rolls a 3, and lose \$2.50 if he rolls any other number, the expected value is $\left(5 \cdot \frac{2}{6}\right) + \left(-1 \cdot \frac{1}{6}\right) + \left(-2.50 \cdot \frac{3}{6}\right)$, which equals 0.25. Thus, he can expect to win \$0.25 after one roll and $(50 \cdot \$0.25)$, or \$12.50, after 50 rolls.

Practice Test #2

Practice Questions

1. Which data set is represented by the boxplot shown below?

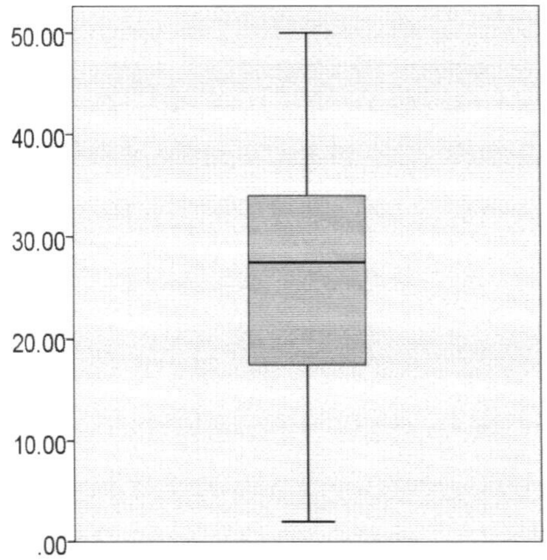

a. 2, 3, 6, 9, 15, 20, 24, 25, 26, 27, 28, 29, 30, 30, 33, 35, 37, 39, 46, 50
b. 1, 6, 6, 7, 10, 14, 14, 19, 20, 21, 24, 28, 35, 37, 38, 39, 41, 44, 44, 45
c. 3, 3, 9, 10, 11, 12, 19, 20, 22, 23, 24, 24, 25, 28, 28, 30, 33, 36, 38, 45
d. 3, 4, 4, 4, 5, 6, 10, 11, 15, 16, 18, 23, 25, 28, 28, 31, 35, 36, 37, 45

2. Given the stem-and-leaf plot shown below, which of the following statements is true?

Data Set A	Stem	Data Set B
1, 1, 8	0	2, 2, 4, 8, 9
4, 5, 5, 8, 9	1	3, 5
0, 1, 4, 7, 8, 9	2	3, 9
	3	1, 6
	4	0, 0, 2
0	5	0

a. Data set A shows a larger center and more variability.
b. Data set A shows a larger center and less variability.
c. Data set A shows a smaller center and more variability.
d. Data set A shows a smaller center and less variability.

3. Given the double bar graph shown below, which of the following statements is true?

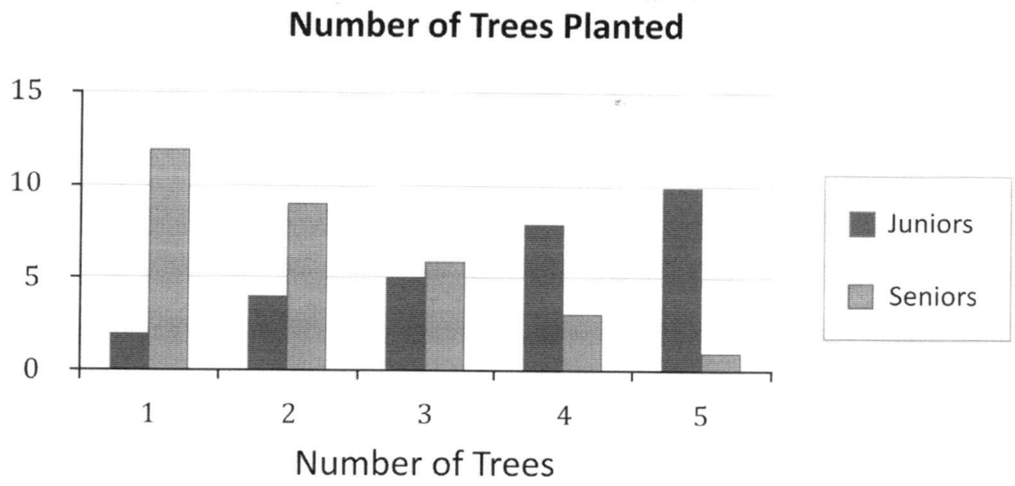

a. The number of trees planted by the Juniors is positively skewed, while the number of trees planted by the Seniors is approximately normal.
b. The number of trees planted by the Juniors is negatively skewed, while the number of trees planted by the Seniors is positively skewed.
c. The number of trees planted by the Juniors is positively skewed, while the number of trees planted by the Seniors is negatively skewed.
d. The number of trees planted by the Juniors is approximately normal, while the number of trees planted by the Juniors is positively skewed.

4. The numbers of volunteers in different states are shown in the table below. Which of the following statements is true?

Texas	8	17	18	19	20	21	21	21	22	28	29	31	41	45	48
New Mexico	7	11	15	29	30	30	31	33	34	36	37	42	44	44	45

a. The numbers of volunteers in Texas have a greater median and mean.
b. The numbers of volunteers in Texas have a greater median and a smaller mean.
c. The numbers of volunteers in Texas have a smaller median and mean.
d. The numbers of volunteers in Texas have a smaller median and a greater mean.

5. Which frequency table is represented by the histogram shown below?

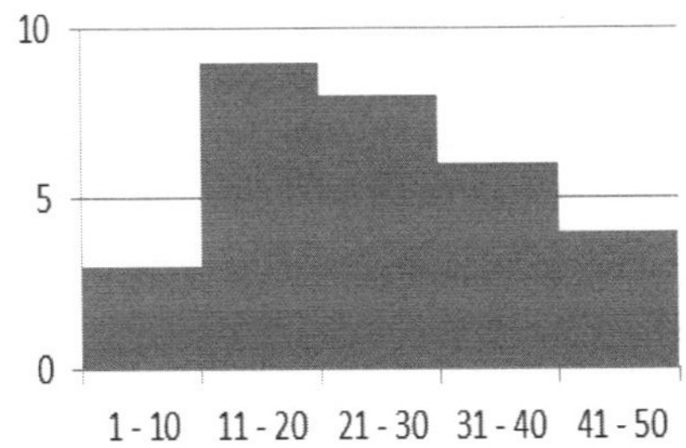

a.

Interval	Frequency
1 – 10	3
11 – 20	9
21 – 30	8
31 – 40	6
41 – 50	4

b.

Interval	Frequency
1 – 10	3
11 – 20	8
21 – 30	8
31 – 40	6
41 – 50	3

c.

Interval	Frequency
1 – 10	3
11 – 20	9
21 – 30	8
31 – 40	5
41 – 50	4

d.

Interval	Frequency
1 – 10	2
11 – 20	9
21 – 30	8
31 – 40	6
41 – 50	4

6. Which of the following measures is unaffected by a skewed distribution?

a. mean
b. median
c. standard deviation
d. range

7. Suppose data are normally distributed. Which of the following best represents the area under the curve, to the right of a z-score of 2.18?

a. 0.01
b. 0.03
c. 0.04
d. 0.06

8. Suppose data are normally distributed. Which of the following best represents the area under the curve, to the left of a z-score of 1.6?

a. 0.87
b. 0.90
c. 0.92
d. 0.95

9. Which of the following represents the area under the curve of a normal distribution, falling between the z-scores of 1.3 and 2.9?

a. 0.09
b. 0.21
c. 0.63
d. 0.90

10. Given the table shown below, what is the probability that a student is a Democrat or prefers math?

	History	Science	Math	Total
Democrat	13	30	37	80
Republican	20	38	12	70
Total	33	68	49	150

a. $\frac{129}{150}$
b. $\frac{46}{75}$
c. $\frac{37}{150}$
d. $\frac{37}{80}$

11. Which of the following best represents the least-squares regression line for the data shown below?

x	1	1	2	3	3	4	5	8	8	9	9	9
y	50	53	61	80	89	85	68	52	64	59	84	100

a. $y = 0.8x + 58.4$
b. $y = 2.2x + 69.2$
c. $y = 1.8x + 44.8$
d. $y = 1.4x + 63.2$

12. Which of the following best represents the margin of error, represented by the residual plot shown below?

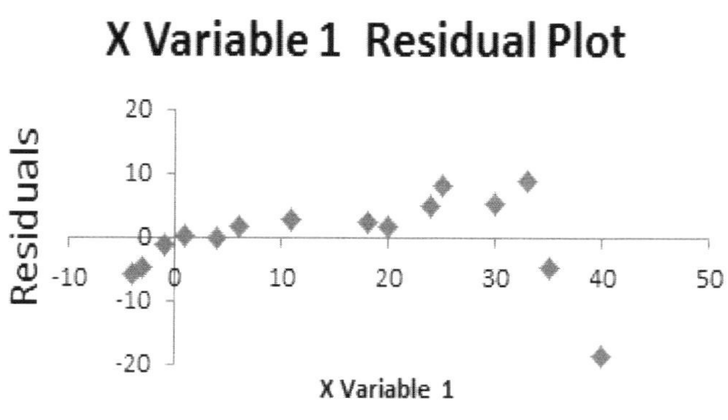

a. 3
b. 7
c. 12
d. 20

13. Given the data set shown below, which of the following best represents the y-intercept of the line of best fit?

x	−4	−3	−2	−1	0	1	2	3	4	5	6	7
y	18	15	10	8	3	2	−2	−6	−10	−10	−16	−15

a. −3.5
b. 8
c. −6
d. 4.5

14. Ana's monthly consulting income may be represented by a line of best fit, revealing constant earnings of $1,500 per month, plus $30 per hour worked. How much can she expect to earn if she works 145 hours during a month?

a. $5,550
b. $6,025
c. $5,850
d. $4,575

15. Given the data shown below, which of the following best represents the value of the correlation coefficient, *r*?

x	2	2	2	3	3	4	4	5	6	6	7	8	9	10	10
y	9	12	21	3	8	18	22	6	12	29	15	8	21	12	28

a. 0.1
b. 0.3
c. 0.5
d. 0.7

16. Which of the following correlation coefficients represents a weak correlation?

a. −0.9
b. 0.2
c. −0.7
d. 0.8

17. This week, Michael drove 218 miles. The average number of weekly miles driven by sales professionals is 225 with a standard deviation of 2 miles. Which of the following statements is true?

a. With a p-value greater than 0.05, the number of miles he drove is significantly different from the average number of weekly miles driven.
b. With a p-value less than 0.05, the number of miles he drove is significantly different from the average number of weekly miles driven.
c. With a p-value greater than 0.05, the number of miles he drove is not significantly different from the average number of weekly miles driven.
d. With a p-value less than 0.05,the number of miles he drove is not significantly different from the average number of weekly miles driven.

18. Jacob's Deli claims to fill each cup of soup with 8 full ounces of soup. In a random sample of 20 cups of soup, the mean number of ounces contained in each cup was 7.7 with a standard deviation of 0.1 ounces. Which of the following statements is true?

a. With a t-value of approximately −10.3, the mean number of ounces found in the random sample of cups is significantly lower.
b. With a t-value of approximately −13.4, the mean number of ounce found in the random sample of cups is significantly lower.
c. With a t-value of approximately −10.3, the mean number of ounces found in the random sample of cups is not significantly lower.
d. With a t-value of approximately −13.4, the mean number of ounces found in the random sample of cups is not significantly lower.

19. Portia tosses a coin 1,000 times. Which of the following best represents the number of times she can expect to get tails?

a. 350
b. 400
c. 450
d. 500

20. A population has a mean of 21 and standard deviation of 2. A sample of 18 subjects reveals a mean of 18. Which of the following statements is true?

a. There is a statistically significant difference between the sample mean and population mean as evidenced by a z-value of approximately −6.4.
b. There is a statistically significant difference between the sample mean and population mean as evidenced by a z-value of approximately −1.2.
c. There is not a statistically significant difference between the sample mean and population mean as evidenced by a z-value of approximately −6.4.
d. There is not a statistically significant difference between the sample mean and population mean as evidenced by a *z*-value of approximately −1.2.

21. An insurance company claims that sales personnel average $42,000 per year with a standard deviation of $1,800. A random sample of 100 sales personnel shows an average of $40,000 per year. Which of the following statements is true?

a. There is a significant difference between the company's annual salary claim and the salary revealed by the sample as evidenced by a p-value less than 0.05.
b. There is a significant difference between the company's annual salary claim and the salary revealed by the sample as evidenced by a p-value greater than 0.05.
c. There is not a significant difference between the company's annual salary claim and the salary revealed by the sample as evidenced by a p-value less than 0.05.
d. There is not a significant difference between the company's annual salary claim and the salary revealed by the sample as evidenced by a *p*-value greater than 0.05.

22. A researcher divides a population into groups of undergraduate students and graduate students and then creates a random sample from each group. Which of the following random sampling techniques does the researcher use?

a. Simple random sampling
b. Stratified random sampling
c. Cluster sampling
d. Systematic sampling

23. When the sampling distribution of means is plotted, which of the following is true?

a. The distribution is approximately normal.
b. The distribution is positively skewed.
c. The distribution is negatively skewed.
d. There is no predictive shape to the distribution.

24. With a 95% level of confidence, which of the following best represents the margin of error for a sample of 60 participants with a mean of 120 and a standard deviation of 10?

a. 1.4
b. 1.9
c. 2.5
d. 3.2

25. With a 95% level of confidence, which of the following best represents the confidence interval for a sample of 100 subjects with a mean of 50 and standard deviation of 2?

a. (48.296, 49.384)
b. (48.792, 49.208)
c. (47.822, 48.964)
d. (49.608, 50.392)

26. Focus group A has 40 members. This group shows a mean of 120 on an assessment with a standard deviation of 15. Focus group B has 45 members. Focus group B has a mean of 130 with a standard deviation of 20. Which of the following statements is true?

a. The difference between the group means is significant as evidenced by a z-score of approximately −1.8.
b. The difference between the group means is significant as evidenced by a z-score of approximately −2.6.
c. The difference between the group means is not significant as evidenced by a z-score of approximately −1.8.
d. The difference between the group means is not significant as evidenced by a *z*-score of approximately −2.6.

27. Elisha spins a spinner with 8 equally-spaced sections labeled 1 – 8. What is the probability that the spinner lands on either a number greater than 5 or on 2?

a. $\frac{3}{64}$
b. $\frac{1}{2}$
c. $\frac{1}{8}$
d. $\frac{3}{8}$

28. Mara rolls a die. What is the probability she rolls a prime number or a 6?

a. $\frac{1}{4}$
b. $\frac{1}{2}$
c. $\frac{3}{4}$
d. $\frac{2}{3}$

29. Abram rolls a die. What is the probability he rolls an even number or a number greater than 4?

a. $\frac{2}{3}$
b. $\frac{1}{6}$
c. $\frac{3}{4}$
d. $\frac{5}{6}$

30. What is the probability Carl draws a heart, replaces it, and then draws another heart?

a. $\frac{1}{2}$
b. $\frac{1}{4}$
c. $\frac{1}{8}$
d. $\frac{1}{16}$

31. What is the probability that Ben draws a 4, does not replace it, and then draws a face card (a jack, queen, or king)?

a. $\frac{12}{663}$
b. $\frac{3}{16}$
c. $\frac{1}{2}$
d. $\frac{3}{169}$

32. A bag contains 2 red marbles, 3 blue marbles, and 5 green marbles. What is the probability Gitta draws a red marble, does not replace it, and then draws another red marble?

a. $\frac{1}{25}$
b. $\frac{1}{45}$
c. $\frac{2}{45}$
d. $\frac{14}{45}$

33. The probability that Jason chooses Mexican food is 0.28. The probability that he chooses a mystery novel is 0.36. The probability that he chooses Mexican food, given that he chooses a mystery novel, is 0.44. Which of the following best represents the probability he chooses Mexican food or a mystery novel?

a. 34%
b. 40%
c. 42%
d. 48%

34. The probability that Robert chooses New Mexico State University is 0.35. The probability that he chooses New Mexico State University or a membership in Kappa Delta Pi is 0.65. The probability that he chooses New Mexico State University and a membership in Kappa Delta Pi is 0.12. What is the probability that he chooses a membership in Kappa Delta Pi?

a. 0.24
b. 0.28
c. 0.38
d. 0.42

35. Given the two-way frequency table shown below, what is the probability that a student is a Sophomore or prefers Lattes?

	Cappuccino	Latte	Espresso	Total
Freshman	20	29	22	71
Sophomore	19	26	22	67
Total	39	55	44	138

a. $\frac{48}{59}$
b. $\frac{37}{51}$
c. $\frac{48}{69}$
d. $\frac{39}{50}$

36. Given the two-way frequency table shown below, which of the following best represents the probability a Representative prefers jazz or is a Republican?

	Classical	Jazz	Rock	Total
Democrat	16	30	28	74
Republican	26	16	27	69
Total	42	46	55	143

a. 57%
b. 62%
c. 69%
d. 73%

37. Martin rolls a die 10 times. Which of the following best represents the probability he rolls exactly three 5's?

a. 0.12
b. 0.16
c. 0.20
d. 0.24

38. Amanda spins a spinner with 8 equally-spaced sections labeled 1 – 8 a total of 6 times. Which of the following best represents the probability that the spinner lands on 7 exactly two times?

a. 0.08
b. 0.12
c. 0.14
d. 0.18

39. Monique draws a number from a bag containing numbers 1 – 20. Each time she draws a number, she reads it and then puts it back in the bag. If she draws 4 numbers from the bag, which of the following best represents the probability that she gets exactly two 8's?

a. 0.01
b. 0.05
c. 0.12
d. 0.16

40. What is the expected value of the number of 4's Olivia will get after rolling a die 10 times?

a. $\frac{3}{4}$
b. $\frac{5}{2}$
c. $\frac{5}{3}$
d. $\frac{8}{9}$

41. What is the expected value of the number of 2's Carlos will get after spinning a spinner four times if the spinner has equal sections labeled 1 – 9?

a. $\frac{2}{9}$
b. $\frac{3}{8}$
c. $\frac{4}{9}$
d. $\frac{1}{3}$

42. What is the expected value of spinning a spinner with 10 equally-spaced sections labeled 1 – 10?

a. 4.5
b. 5
c. 5.5
d. 6

43. Mary is asked to roll a standard die. She will receive $2 if she rolls a 3, $4 if she rolls a 4, and $6 if she rolls a 6. What is the expected value for the amount of money she will receive?

a. $2
b. $3
c. $4
d. $5

44. Todd will win $10 if he rolls a 3 or 4, $8 if he rolls a 2, and $12 if he rolls any other number. Which of the following best represents the amount of money he can expect to win?

a. $9.37
b. $9.67
c. $10.67
d. $11.37

45. Given the ordered pairs below, in the form, $(x, f(x))$, which represent values and their corresponding probabilities, what is the expected value?

$$(0, 18), (1, 12), (2, 20), (3, 24), (4, 10), (5, 36), (6, 4)$$

a. 354
b. 362
c. 368
d. 374

46. Penny draws 2 cards from a standard deck of cards. What is the expected value of drawing a club?

a. 0.25
b. 0.50
c. 0.75
d. 1

47. You draw two cards from a deck with cards labeled 1 – 6 and then graph the probabilities of the expected value of picking a 3 after choosing 2 cards. Which of the following represents the probabilities that correspond to x-values of 0, 1, and 2 cards?

a. $\frac{2}{3}, \frac{1}{3}, 0$
b. $\frac{1}{3}, \frac{1}{6}, 0$
c. $\frac{2}{3}, \frac{1}{3}, \frac{1}{5}$
d. $\frac{1}{3}, \frac{1}{6}, \frac{1}{9}$

48. Amanda rolls a die. She will lose \$6 if the die lands on a 3 or 5. She will win \$3 if the die lands on a 1 or 2. She will lose \$9 if the die lands on a 4 or 6. What is the expected value?

a. –\$2
b. –\$4
c. –\$5
d. –\$6

49. Marlon rolls a die. He will win \$45 if the die lands on a number greater than 2 and lose \$30 if the die lands on any other number. What is the expected value of his earnings?

a. \$18
b. \$20
c. \$22
d. \$24

50. Aubrey spins a spinner with 8 equally-spaced sections labeled 1 – 8. She will lose \$90 if the spinner lands on a 1, 5, or 8. She will win \$100 if the spinner lands on any other number. What is the expected value?

a. \$27.25
b. \$27.75
c. \$28.25
d. \$28.75

Answers and Explanations

Math.Content.S.ID.1

1. A: The data shown for Choice A has a minimum of 2, maximum of 50, median of 27.5, first quartile of 17.5, and third quartile of 34. The given boxplot represents each of these values.

Math.Content.S.ID.2

2. D: Data set B has a larger center because there are more high scores. Data set A shows less variability, due to the clustering of scores between 1 and 29.

Math.Content.S.ID.3

3. B: The number of trees planted by the Juniors is skewed left, with more higher numbers of trees planted. The number of trees planted by the Seniors is skewed right, with more individuals with lower numbers of trees planted.

Math.Content.S.ID.1

4. C: The median number of volunteers in Texas is 21, with a mean of approximately 25.9. The median number of volunteers in New Mexico is 33, with a mean of approximately 31.2. Thus, the median and mean numbers of volunteers in Texas are smaller.

Math.Content.S.ID.3

5. A: The frequency table for Choice A correctly shows the frequencies represented by the histogram. The frequencies of values, falling between 1 and 10 is 3, between 11 and 20 is 9, between 21 and 30 is 8, between 31 and 40 is 6, and between 41 and 50 is 4.

Math.Content.S.ID.1

6. B: The median is unaffected by a skewed distribution (i.e., by extreme outliers). The median represents the value that 50 percent of the scores fall above and 50 percent of the scores fall below. The mean, standard deviation, and range are all impacted by non-normal data.

Math.Content.S.ID.4

7. A: Using a standard z-table, the area from the mean to z is shown to be 0.4854. This area should be subtracted from 0.5 in order to show the area to the right of a z-score of 2.18. Thus, the area to the right of the z-score is 0.0146, or approximately 0.01. A z-table, including the smaller portion of area under a curve may also be used. This area is reported as 0.0146.

Math.Content.S.ID.4

8. D: Using a standard z-table, the area from the mean to z is shown to be 0.4452. This area should be added to 0.5 in order to show the area to the left of a z-score of 1.6. Thus, the area to the left of the z-score is 0.9452, or approximately 0.95. A z-table, including the larger portion of area under a curve may also be used. This area is reported as 0.9452.

Math.Content.S.ID.4

9. A: A standard z-table shows an area of 0.4032 between the mean and a z-score of 1.3. It also shows an area of 0.4981 between the mean and a z-score of 2.9. The difference of the two areas is 0.0949. Thus, the area between the two z-scores is approximately 9%. Using a table that reports the mean to z, the mean-to-z area for a z-score of 1.3 may be subtracted from the mean-to-z area for a z-score of 2.9.

Math.Content.S.ID.5

10. B: The probability may be written as $P(A \text{ or } B) = P(A) + P(B) - P(A \text{ and } B) = \frac{80}{150} + \frac{49}{150} - \frac{37}{150}$, which simplifies to $P(A \text{ or } B) = \frac{46}{75}$.

Math.Content.S.ID.6a

11. D: The least-squares regression line may be found by entering the x- and y-values as lists into the graphing calculator. Choose Stat, Calc, and then LinReg. The least-squares regression line is represented by the form, $y = ax + b$, where a represents the slope and b represents the y-intercept. Thus, the best estimate for the least-squares regression line is $y = 1.4x + 63.2$.

Math.Content.S.ID.6b

12. B: The values deviate between + or −10 points from the regression line. The margin of error of 7 best represents the average of these deviations.

Math.Content.S.ID.7

13. D: The least-squares regression line may be found by entering the x- and y-values as lists into the graphing calculator. Choose Stat, Calc, and then LinReg. The least-squares regression line is represented by the form, $y = ax + b$, where a represents the slope and b represents the y-intercept. Thus, the best estimate for the y-intercept is 4.5.

Math.Content.S.ID.9

14. C: The line of best fit may be written as $y = 30x + 1500$, where x represents the number of hours worked. Substituting 145 for x gives $y = 30(145) + 1500$, which equals 5,850. Thus, she can expect to earn $5,850 during a month, in which she works 145 hours.

Math.Content.S.ID.8

15. B: The value of the correlation coefficient, r, may be found by using the LinReg feature of a graphing calculator. Enter the x- and y-values as two separate lists, and then choose Stat, Calc, and LinReg. The resulting r-value is the correlation coefficient. Thus, the correlation coefficient, r, for this data is approximately 0.3.

Math.Content.S.ID.8

16. B: Correlation coefficients range from −1 to 1, with values close to −1 or 1, representing strong relationships. Thus, a correlation coefficient of 0.2 represents a weak correlation.

Math.Content.S.IC.1

17. B: The z-score may be calculated, using the formula, $z = \frac{X-\mu}{\sigma}$. Substituting a value of 218 for X, 225 for μ, and 2 for σ gives $z = \frac{218-225}{2}$ or $z = -3.5$. A z-score of 3.5 has a p-value of 0.0013, which is smaller than 0.05. Therefore, there is a statistically significant difference between the number of miles he drove during the week and the average number of weekly miles driven.

Math.Content.S.IC.1

18. B: Since only the sample standard deviation is known, a t-distribution will be used. When comparing a sample mean to a population mean and using a t-distribution, use the following formula: $t = \frac{\bar{X}-\mu}{\frac{s}{\sqrt{n}}}$ where $\bar{X}$ represents the sample mean, μ represents the population mean, s represents the sample standard deviation, and n represents the sample size. Substituting 7.7 for $\bar{X}$, 8 for μ, 0.1 for s, and 20 for n gives $t = \frac{7.7-8}{\frac{0.1}{\sqrt{20}}}$, or $t \approx -13.4$. The absolute value of −13.4, namely 13.4, is more than the critical t-value of approximately 1.729. Note that the critical t-value for a one-tailed test with a 0.05 level of significance for 19 degrees of freedom was used. Since the absolute value of the actual t-value is more than the critical t-value, the mean of the random sample of cups is interpreted as significantly lower.

Math.Content.S.IC2

19. D: The theoretical probability of getting tails is $\frac{1}{2}$. Thus, she can expect to get a total of $\frac{1}{2} \times 1{,}000$ tails, or 500 tails.

Math.Content.S.IC.1

20. A: A z-distribution will be used since the population standard deviation is known. When comparing a sample mean to a population mean, the following formula should be used: $z = \frac{\bar{X}-\mu}{\frac{\sigma}{\sqrt{n}}}$, where $\bar{X}$ represents the sample mean, μ represents the population mean, s represents the population standard deviation, and n represents the sample size. Substituting 18 for $\bar{X}$, 21 for μ, 2 for σ, and 18 for n gives $z = \frac{18-21}{\frac{2}{\sqrt{18}}}$, or $z \approx -6.4$. The p-value for a z-score of 6.4 is less than 0.001. Therefore, there is a significant difference between the means.

Math.Content.S.IC.1

21. A: A z-distribution will be used since the population standard deviation is known. When comparing a sample mean to a population mean, use the formula, $z = \frac{\bar{X}-\mu}{\frac{\sigma}{\sqrt{n}}}$, where $\bar{X}$ represents the sample mean, μ represents the population mean, σ represents the population standard deviation, and n represents the sample size. Substituting 40,000 for $\bar{X}$, 42,000 for μ, 1,800 for σ, and 100 for n gives $z = \frac{40{,}000-42{,}000}{\frac{1800}{\sqrt{100}}}$, or $z \approx -11.1$. The p-value for a z-score of 11.1 is less than 0.001. Therefore, there is a significant difference between the means.

Math.Content.S.IC.3

22. C: When first dividing groups into clusters and then completing random samples within the groups, cluster sampling is used. In this case, the clusters are undergraduate students and graduate students.

Math.Content.S.ID

23. A: When plotting the sampling distribution of means, the distribution is approximately normal. As the number of random samples increases, the plotting of the means approaches a normal distribution.

Math.Content.S.IC.6

24. C: The formula for finding the margin of error is $E = z_{\alpha/2} \cdot \frac{\sigma}{\sqrt{n}}$, where z represents the z-score for a 95% degree of confidence for a two-tailed test, σ represents the standard deviation, and n represents the sample size. Thus, the margin of error may be represented as $E = 1.96 \cdot \frac{10}{\sqrt{60}}$, which is approximately 2.5.

Math.Content.S.IC.4

25. D: The margin of error may be represented as $E = 1.96 \cdot \frac{2}{\sqrt{100}}$, or approximately 0.392. Subtracting 0.392 from 50 gives 49.608. Adding 0.392 to 50 gives 50.392. Thus, the confidence interval is (49.608, 50.392).

Math.Content.S.IC.5

26. B: The problem asks you to compare two means. A two-sample z-test may be used. The given means, standard deviations, and sample sizes may be entered into a graphing calculator, in order to determine the critical z-value and p-value. Choose Stat, Tests, and 2-SampZTest. The resulting z-score is approximately −2.6 with a p-value of approximately 0.01. Therefore, there is a significant difference between the means of the focus groups.

Math.Content.S.CP.7

27. B: The probability of mutually exclusive events, A or B, occurring may be written as $P(A \text{ or } B) = P(A) + P(B)$. Thus, $P(A \text{ or } B) = \frac{3}{8} + \frac{1}{8} \text{ or } \frac{1}{2}$.

Math.Content.S.CP.7

28. D: The probability of mutually exclusive events, A or B, occurring may be written as $P(A \text{ or } B) = P(A) + P(B)$. Thus, $P(A \text{ or } B) = \frac{3}{6} + \frac{1}{6} \text{ or } \frac{2}{3}$.

Math.Content.S.CP.1

29. A: The probability of non-mutually exclusive events, A or B, occurring may be written as $P(A \text{ or } B) = P(A) + P(B) - P(A \text{ and } B)$. Thus, $P(A \text{ or } B) = \frac{3}{6} + \frac{2}{6} - \frac{1}{6} \text{ or } \frac{2}{3}$.

Math.Content.S.CP.2

30. D: The probability of independent events, A and B, may be found using the formula, $P(A \text{ and } B) = P(A) \cdot P(B)$. Thus, the probability he draws a heart, replaces it, and then draws another heart may be represented as $P(A \text{ and } B) = \frac{13}{52} \cdot \frac{13}{52}$, which simplifies to $P(A \text{ and } B) = \frac{1}{16}$.

Math.Content.S.CP.3

31. A: Since he does not replace the first card, the events are dependent. The sample space will decrease by 1 for the second draw because there will be one less card to choose from. Thus, the probability may be written as $P(A \text{ and } B) = \frac{13}{52} \times \frac{13}{51}$, or $P(A \text{ and } B) = \frac{169}{2{,}652}$, which simplifies to $\frac{13}{204}$.

Math.Content.S.CP.6

32. B: The events are dependent since the first marble was not replaced. The sample space of the second draw will decrease by 1 because there will be one fewer marble to choose from. The number of possible red marbles for the second draw will also decrease by 1. Thus, the probability may be written as $P(A \text{ and } B) = \frac{2}{10} \times \frac{1}{9}$. Thus, the probability Gitta draws a red marble, does not replace it, and draws another red marble is $\frac{2}{90}$, or $\frac{1}{45}$.

Math.Content.S.CP.8

33. D: The probability of events, A or B, occurring may be found, using the formula, $P(A \text{ or } B) = P(A) + P(B) - P(A \text{ and } B)$. To find $P(A \text{ and } B)$, use the formula $P(A \text{ and } B) = P(B) \cdot P(A|B)$. Thus, $P(A \text{ and } B) = 0.36 \cdot 0.44$, or 0.1584. The probabilities may now be substituted, giving $P(A \text{ or } B) = 0.28 + 0.36 - 0.1584$. Thus, the probability that he chooses Mexican food or a mystery novel is approximately 48%.

Math.Content.S.CP.8

34. D: The probability of events, A or B, occurring may be found using the formula, $P(A \text{ or } B) = P(A) + P(B) - P(A \text{ and } B)$. The given probabilities may be substituted into the formula by writing $0.65 = 0.35 + P(B) - 0.12$. Solving for P(B) gives a probability of 0.42. Therefore, the probability that he chooses a membership in Kappa Delta Pi is 0.42, or 42%.

Math.Content.S.CP.4

35. C: The probability may be written as: $P(S \text{ or } L) = \frac{67}{138} + \frac{55}{138} - \frac{26}{138}$. Thus, the probability a student is a Sophomore or prefers Lattes is $\frac{48}{69}$.

Math.Content.S.CP.5

36. C: The probability may be written as: $P(J \text{ or } R) = \frac{46}{143} + \frac{69}{143} - \frac{16}{143}$. Thus, the probability a Representative prefers jazz or is a Republican is approximately 69%.

Math.Content.S.CP.9

37. B: The binomial pdf function of a graphing calculator may be used to compute the probability. The following may be entered into the graphing calculator: binompdf$\left(10, \frac{1}{6}, 3\right)$, which is approximately 0.16.

Math.Content.S.CP.9

38. C: The following may be entered into the graphing calculator: binompdf$\left(6, \frac{1}{8}, 2\right)$, which is approximately 0.14.

Math.Content.S.CP.9

39. A: The following may be entered into the graphing calculator: binompd$\left(4, \frac{1}{20}, 2\right)$, which is approximately 0.01.

Math.Content.S.MD.2

40. C: The expected value is equal to the product of the theoretical probability of getting a 4 after one roll, and the number of rolls, 10. Thus, the expected value is $\frac{1}{6} \cdot 10$, or $\frac{5}{3}$.

Math.Content.S.MD.2

41. C: The expected value is equal to the product of the theoretical probability of getting a 2 after one spin and the number of spins, 4. Thus, the expected value is $\frac{1}{9} \cdot 4$, or $\frac{4}{9}$.

Math.Content.S.MD.2

42. C: The expected value is equal to the sum of the products of the probability of rolling each number and the number's value. Thus, the expected value is $\left(1 \cdot \frac{1}{10}\right) + \left(2 \cdot \frac{1}{10}\right) + \left(3 \cdot \frac{1}{10}\right) + \left(4 \cdot \frac{1}{10}\right) + \left(5 \cdot \frac{1}{10}\right) + \left(6 \cdot \frac{1}{10}\right) + \left(7 \cdot \frac{1}{10}\right) + \left(8 \cdot \frac{1}{10}\right) + \left(9 \cdot \frac{1}{10}\right) + \left(10 \cdot \frac{1}{10}\right)$. The expected value is 5.5.

Math.Content.S.MD.4

43. A: The expected value is equal to the sum of the products of the probability of rolling each of the numbers and the amount she will win for each number. Thus, the expected value is $\left(2 \cdot \frac{1}{6}\right) + \left(4 \cdot \frac{1}{6}\right) + \left(6 \cdot \frac{1}{6}\right)$, which equal 2. She can expect to win $2.

Math.Content.S.MD.5a

44. C: The expected value is equal to the sum of the products of the probabilities and the amount he will win. The probability of rolling a 3 or 4 is $\frac{2}{6}$, the probability of rolling a 2 is $\frac{1}{6}$, and the probability of rolling any other number is $\frac{3}{6}$. Thus, the expected value is $\left(10 \cdot \frac{2}{6}\right) + \left(8 \cdot \frac{1}{6}\right) + \left(12 \cdot \frac{3}{6}\right)$. Thus, he can expect to win approximately $10.67.

Math.Content.S.MD.1

45. C: The expected value is equal to the sum of the products of the probabilities and their x-values. Thus, the expected value is $(0 \cdot 18) + (1 \cdot 12) + (2 \cdot 20) + (3 \cdot 24) + (4 \cdot 10) + (5 \cdot 36) + (6 \cdot 4)$. The expected value equals 368.

Math.Content.S.MD.3

46. B: The table below may be used to calculate the probabilities:

x	f(x)
0	$\frac{\binom{13}{0}\binom{39}{2}}{\binom{52}{2}}$
1	$\frac{\binom{13}{1}\binom{39}{1}}{\binom{52}{2}}$
2	$\frac{\binom{13}{2}\binom{39}{0}}{\binom{52}{2}}$

The expected value is $\left(0 \cdot \frac{741}{1326}\right) + \left(1 \cdot \frac{507}{1326}\right) + \left(2 \cdot \frac{78}{1326}\right)$, which is equal to 0.5.

Math.Content.S.MD.1 and S.MD.7

47. A: The table below represents the probabilities:

x	f(x)
0	$\frac{\binom{1}{0}\binom{5}{2}}{\binom{6}{2}} = \frac{2}{3}$
1	$\frac{\binom{1}{1}\binom{5}{1}}{\binom{6}{2}} = \frac{1}{3}$
2	$\frac{\binom{1}{2}\binom{5}{0}}{\binom{6}{2}} = 0$

Math.Content.S.MD.5b and S.MD.6

48. B: The expected value is equal to the sum of the products of the probabilities and the amount she will lose or win. Each probability is equal to $\frac{2}{6}$. Thus, the expected value may be written as $\left(-6 \cdot \frac{2}{6}\right) + \left(3 \cdot \frac{2}{6}\right) + \left(-9 \cdot \frac{2}{6}\right)$. Thus, the expected value is –\$4.

Math.Content.S.MD.5b and S.MD.6

49. B: The expected value is equal to the sum of the products of the probabilities and the amount he will lose or win. The probability of rolling a number greater than 2 is $\frac{4}{6}$, and the probability of rolling another number is $\frac{2}{6}$. Thus, the expected value is $\left(45 \cdot \frac{4}{6}\right) + \left(-30 \cdot \frac{2}{6}\right)$. Thus, the expected value is \$20.

Math.Content.S.MD.5b and S.MD.6

50. D: The expected value is equal to the sum of the products of the probabilities and the amount she will lose or win. The probability the spinner will land on a 1, 5, or 8 is $\frac{3}{8}$. The probability of getting another number is $\frac{5}{8}$. Thus, the expected value is $\left(-90 \cdot \frac{3}{8}\right) + \left(100 \cdot \frac{5}{8}\right)$. The expected value is \$28.75.

How to Overcome Test Anxiety

Just the thought of taking a test is enough to make most people a little nervous. A test is an important event that can have a long-term impact on your future, so it's important to take it seriously and it's natural to feel anxious about performing well. But just because anxiety is normal, that doesn't mean that it's helpful in test taking, or that you should simply accept it as part of your life. Anxiety can have a variety of effects. These effects can be mild, like making you feel slightly nervous, or severe, like blocking your ability to focus or remember even a simple detail.

If you experience test anxiety—whether severe or mild—it's important to know how to beat it. To discover this, first you need to understand what causes test anxiety.

Causes of Test Anxiety

While we often think of anxiety as an uncontrollable emotional state, it can actually be caused by simple, practical things. One of the most common causes of test anxiety is that a person does not feel adequately prepared for their test. This feeling can be the result of many different issues such as poor study habits or lack of organization, but the most common culprit is time management. Starting to study too late, failing to organize your study time to cover all of the material, or being distracted while you study will mean that you're not well prepared for the test. This may lead to cramming the night before, which will cause you to be physically and mentally exhausted for the test. Poor time management also contributes to feelings of stress, fear, and hopelessness as you realize you are not well prepared but don't know what to do about it.

Other times, test anxiety is not related to your preparation for the test but comes from unresolved fear. This may be a past failure on a test, or poor performance on tests in general. It may come from comparing yourself to others who seem to be performing better or from the stress of living up to expectations. Anxiety may be driven by fears of the future—how failure on this test would affect your educational and career goals. These fears are often completely irrational, but they can still negatively impact your test performance.

Elements of Test Anxiety

As mentioned earlier, test anxiety is considered to be an emotional state, but it has physical and mental components as well. Sometimes you may not even realize that you are suffering from test anxiety until you notice the physical symptoms. These can include trembling hands, rapid heartbeat, sweating, nausea, and tense muscles. Extreme anxiety may lead to fainting or vomiting. Obviously, any of these symptoms can have a negative impact on testing. It is important to recognize them as soon as they begin to occur so that you can address the problem before it damages your performance.

The mental components of test anxiety include trouble focusing and inability to remember learned information. During a test, your mind is on high alert, which can help you recall information and stay focused for an extended period of time. However, anxiety interferes with your mind's natural processes, causing you to blank out, even on the questions you know well. The strain of testing during anxiety makes it difficult to stay focused, especially on a test that may take several hours. Extreme anxiety can take a huge mental toll, making it difficult not only to recall test information but even to understand the test questions or pull your thoughts together.

Effects of Test Anxiety

Test anxiety is like a disease—if left untreated, it will get progressively worse. Anxiety leads to poor performance, and this reinforces the feelings of fear and failure, which in turn lead to poor performances on subsequent tests. It can grow from a mild nervousness to a crippling condition. If allowed to progress, test anxiety can have a big impact on your schooling, and consequently on your future.

Test anxiety can spread to other parts of your life. Anxiety on tests can become anxiety in any stressful situation, and blanking on a test can turn into panicking in a job situation. But fortunately, you don't have to let anxiety rule your testing and determine your grades. There are a number of relatively simple steps you can take to move past anxiety and function normally on a test and in the rest of life.

Physical Steps for Beating Test Anxiety

While test anxiety is a serious problem, the good news is that it can be overcome. It doesn't have to control your ability to think and remember information. While it may take time, you can begin taking steps today to beat anxiety.

Just as your first hint that you may be struggling with anxiety comes from the physical symptoms, the first step to treating it is also physical. Rest is crucial for having a clear, strong mind. If you are tired, it is much easier to give in to anxiety. But if you establish good sleep habits, your body and mind will be ready to perform optimally, without the strain of exhaustion. Additionally, sleeping well helps you to retain information better, so you're more likely to recall the answers when you see the test questions.

Getting good sleep means more than going to bed on time. It's important to allow your brain time to relax. Take study breaks from time to time so it doesn't get overworked, and don't study right before bed. Take time to rest your mind before trying to rest your body, or you may find it difficult to fall asleep.

Along with sleep, other aspects of physical health are important in preparing for a test. Good nutrition is vital for good brain function. Sugary foods and drinks may give a burst of energy but this burst is followed by a crash, both physically and emotionally. Instead, fuel your body with protein and vitamin-rich foods.

Also, drink plenty of water. Dehydration can lead to headaches and exhaustion, especially if your brain is already under stress from the rigors of the test. Particularly if your test is a long one, drink water during the breaks. And if possible, take an energy-boosting snack to eat between sections.

Along with sleep and diet, a third important part of physical health is exercise. Maintaining a steady workout schedule is helpful, but even taking 5-minute study breaks to walk can help get your blood pumping faster and clear your head. Exercise also releases endorphins, which contribute to a positive feeling and can help combat test anxiety.

When you nurture your physical health, you are also contributing to your mental health. If your body is healthy, your mind is much more likely to be healthy as well. So take time to rest, nourish your body with healthy food and water, and get moving as much as possible. Taking these physical steps will make you stronger and more able to take the mental steps necessary to overcome test anxiety.

Mental Steps for Beating Test Anxiety

Working on the mental side of test anxiety can be more challenging, but as with the physical side, there are clear steps you can take to overcome it. As mentioned earlier, test anxiety often stems from lack of preparation, so the obvious solution is to prepare for the test. Effective studying may be the most important weapon you have for beating test anxiety, but you can and should employ several other mental tools to combat fear.

First, boost your confidence by reminding yourself of past success—tests or projects that you aced. If you're putting as much effort into preparing for this test as you did for those, there's no reason you should expect to fail here. Work hard to prepare; then trust your preparation.

Second, surround yourself with encouraging people. It can be helpful to find a study group, but be sure that the people you're around will encourage a positive attitude. If you spend time with others who are anxious or cynical, this will only contribute to your own anxiety. Look for others who are motivated to study hard from a desire to succeed, not from a fear of failure.

Third, reward yourself. A test is physically and mentally tiring, even without anxiety, and it can be helpful to have something to look forward to. Plan an activity following the test, regardless of the outcome, such as going to a movie or getting ice cream.

When you are taking the test, if you find yourself beginning to feel anxious, remind yourself that you know the material. Visualize successfully completing the test. Then take a few deep, relaxing breaths and return to it. Work through the questions carefully but with confidence, knowing that you are capable of succeeding.

Developing a healthy mental approach to test taking will also aid in other areas of life. Test anxiety affects more than just the actual test—it can be damaging to your mental health and even contribute to depression. It's important to beat test anxiety before it becomes a problem for more than testing.

Study Strategy

Being prepared for the test is necessary to combat anxiety, but what does being prepared look like? You may study for hours on end and still not feel prepared. What you need is a strategy for test prep. The next few pages outline our recommended steps to help you plan out and conquer the challenge of preparation.

Step 1: Scope Out the Test

Learn everything you can about the format (multiple choice, essay, etc.) and what will be on the test. Gather any study materials, course outlines, or sample exams that may be available. Not only will this help you to prepare, but knowing what to expect can help to alleviate test anxiety.

Step 2: Map Out the Material

Look through the textbook or study guide and make note of how many chapters or sections it has. Then divide these over the time you have. For example, if a book has 15 chapters and you have five days to study, you need to cover three chapters each day. Even better, if you have the time, leave an extra day at the end for overall review after you have gone through the material in depth.

If time is limited, you may need to prioritize the material. Look through it and make note of which sections you think you already have a good grasp on, and which need review. While you are studying, skim quickly through the familiar sections and take more time on the challenging parts.

Write out your plan so you don't get lost as you go. Having a written plan also helps you feel more in control of the study, so anxiety is less likely to arise from feeling overwhelmed at the amount to cover.

Step 3: Gather Your Tools

Decide what study method works best for you. Do you prefer to highlight in the book as you study and then go back over the highlighted portions? Or do you type out notes of the important information? Or is it helpful to make flashcards that you can carry with you? Assemble the pens, index cards, highlighters, post-it notes, and any other materials you may need so you won't be distracted by getting up to find things while you study.

If you're having a hard time retaining the information or organizing your notes, experiment with different methods. For example, try color-coding by subject with colored pens, highlighters, or post-it notes. If you learn better by hearing, try recording yourself reading your notes so you can listen while in the car, working out, or simply sitting at your desk. Ask a friend to quiz you from your flashcards, or try teaching someone the material to solidify it in your mind.

Step 4: Create Your Environment

It's important to avoid distractions while you study. This includes both the obvious distractions like visitors and the subtle distractions like an uncomfortable chair (or a too-comfortable couch that makes you want to fall asleep). Set up the best study environment possible: good lighting and a comfortable work area. If background music helps you focus, you may want to turn it on, but otherwise keep the room quiet. If you are using a computer to take notes, be sure you don't have any other windows open, especially applications like social media, games, or anything else that could distract you. Silence your phone and turn off notifications. Be sure to keep water close by so you stay hydrated while you study (but avoid unhealthy drinks and snacks).

Also, take into account the best time of day to study. Are you freshest first thing in the morning? Try to set aside some time then to work through the material. Is your mind clearer in the afternoon or evening? Schedule your study session then. Another method is to study at the same time of day that you will take the test, so that your brain gets used to working on the material at that time and will be ready to focus at test time.

Step 5: Study!

Once you have done all the study preparation, it's time to settle into the actual studying. Sit down, take a few moments to settle your mind so you can focus, and begin to follow your study plan. Don't give in to distractions or let yourself procrastinate. This is your time to prepare so you'll be ready to fearlessly approach the test. Make the most of the time and stay focused.

Of course, you don't want to burn out. If you study too long you may find that you're not retaining the information very well. Take regular study breaks. For example, taking five minutes out of every hour to walk briskly, breathing deeply and swinging your arms, can help your mind stay fresh.

As you get to the end of each chapter or section, it's a good idea to do a quick review. Remind yourself of what you learned and work on any difficult parts. When you feel that you've mastered the material, move on to the next part. At the end of your study session, briefly skim through your notes again.

But while review is helpful, cramming last minute is NOT. If at all possible, work ahead so that you won't need to fit all your study into the last day. Cramming overloads your brain with more information than it can process and retain, and your tired mind may struggle to recall even previously learned information when it is overwhelmed with last-minute study. Also, the urgent nature of cramming and the stress placed on your brain contribute to anxiety. You'll be more likely to go to the test feeling unprepared and having trouble thinking clearly.

So don't cram, and don't stay up late before the test, even just to review your notes at a leisurely pace. Your brain needs rest more than it needs to go over the information again. In fact, plan to finish your studies by noon or early afternoon the day before the test. Give your brain the rest of the day to relax or focus on other things, and get a good night's sleep. Then you will be fresh for the test and better able to recall what you've studied.

Step 6: Take a Practice Test

Many courses offer sample tests, either online or in the study materials. This is an excellent resource to check whether you have mastered the material, as well as to prepare for the test format and environment.

Check the test format ahead of time: the number of questions, the type (multiple choice, free response, etc.), and the time limit. Then create a plan for working through them. For example, if you have 30 minutes to take a 60-question test, your limit is 30 seconds per question. Spend less time on the questions you know well so that you can take more time on the difficult ones.

If you have time to take several practice tests, take the first one open book, with no time limit. Work through the questions at your own pace and make sure you fully understand them. Gradually work up to taking a test under test conditions: sit at a desk with all study materials put away and set a timer. Pace yourself to make sure you finish the test with time to spare and go back to check your answers if you have time.

After each test, check your answers. On the questions you missed, be sure you understand why you missed them. Did you misread the question (tests can use tricky wording)? Did you forget the information? Or was it something you hadn't learned? Go back and study any shaky areas that the practice tests reveal.

Taking these tests not only helps with your grade, but also aids in combating test anxiety. If you're already used to the test conditions, you're less likely to worry about it, and working through tests until you're scoring well gives you a confidence boost. Go through the practice tests until you feel comfortable, and then you can go into the test knowing that you're ready for it.

Test Tips

On test day, you should be confident, knowing that you've prepared well and are ready to answer the questions. But aside from preparation, there are several test day strategies you can employ to maximize your performance.

First, as stated before, get a good night's sleep the night before the test (and for several nights before that, if possible). Go into the test with a fresh, alert mind rather than staying up late to study.

Try not to change too much about your normal routine on the day of the test. It's important to eat a nutritious breakfast, but if you normally don't eat breakfast at all, consider eating just a protein bar. If you're a coffee drinker, go ahead and have your normal coffee. Just make sure you time it so that the caffeine doesn't wear off right in the middle of your test. Avoid sugary beverages, and drink enough water to stay hydrated but not so much that you need a restroom break 10 minutes into the test. If your test isn't first thing in the morning, consider going for a walk or doing a light workout before the test to get your blood flowing.

Allow yourself enough time to get ready, and leave for the test with plenty of time to spare so you won't have the anxiety of scrambling to arrive in time. Another reason to be early is to select a good seat. It's helpful to sit away from doors and windows, which can be distracting. Find a good seat, get out your supplies, and settle your mind before the test begins.

When the test begins, start by going over the instructions carefully, even if you already know what to expect. Make sure you avoid any careless mistakes by following the directions.

Then begin working through the questions, pacing yourself as you've practiced. If you're not sure on an answer, don't spend too much time on it, and don't let it shake your confidence. Either skip it and come back later, or eliminate as many wrong answers as possible and guess among the remaining ones. Don't dwell on these questions as you continue—put them out of your mind and focus on what lies ahead.

Be sure to read all of the answer choices, even if you're sure the first one is the right answer. Sometimes you'll find a better one if you keep reading. But don't second-guess yourself if you do immediately know the answer. Your gut instinct is usually right. Don't let test anxiety rob you of the information you know.

If you have time at the end of the test (and if the test format allows), go back and review your answers. Be cautious about changing any, since your first instinct tends to be correct, but make sure you didn't misread any of the questions or accidentally mark the wrong answer choice. Look over any you skipped and make an educated guess.

At the end, leave the test feeling confident. You've done your best, so don't waste time worrying about your performance or wishing you could change anything. Instead, celebrate the successful completion of this test. And finally, use this test to learn how to deal with anxiety even better next time.

Review Video: Test Anxiety
Visit mometrix.com/academy and enter code: 100340

Important Qualification

Not all anxiety is created equal. If your test anxiety is causing major issues in your life beyond the classroom or testing center, or if you are experiencing troubling physical symptoms related to your anxiety, it may be a sign of a serious physiological or psychological condition. If this sounds like your situation, we strongly encourage you to seek professional help.

How to Overcome Your Fear of Math

Not again. You're sitting in math class, look down at your test, and immediately start to panic. Your stomach is in knots, your heart is racing, and you break out in a cold sweat. You're staring at the paper, but everything looks like it's written in a foreign language. Even though you studied, you're blanking out on how to begin solving these problems.

Does this sound familiar? If so, then you're not alone! You may be like millions of other people who experience math anxiety. Anxiety about performing well in math is a common experience for students of all ages. In this article, we'll discuss what math anxiety is, common misconceptions about learning math, and tips and strategies for overcoming math anxiety.

What Is Math Anxiety?

Psychologist Mark H. Ashcraft explains math anxiety as a feeling of tension, apprehension, or fear that interferes with math performance. Having math anxiety negatively impacts people's beliefs about themselves and what they can achieve. It hinders achievement within the math classroom and affects the successful application of mathematics in the real world.

Symptoms and Signs of Math Anxiety

To overcome math anxiety, you must recognize its symptoms. Becoming aware of the signs of math anxiety is the first step in addressing and resolving these fears.

Negative Self-Talk

If you have math anxiety, you've most likely said at least one of these statements to yourself:

- "I hate math."
- "I'm not good at math."
- "I'm not a math person."

The way we speak to ourselves and think about ourselves matters. Our thoughts become our words, our words become our actions, and our actions become our habits. Thinking negatively about math creates a self-fulfilling prophecy. In other words, if you take an idea as a fact, then it will come true because your behaviors will align to match it.

Avoidance

Some people who are fearful or anxious about math will tend to avoid it altogether. Avoidance can manifest in the following ways:

- Lack of engagement with math content
- Not completing homework and other assignments
- Not asking for help when needed
- Skipping class
- Avoiding math-related courses and activities

Avoidance is one of the most harmful impacts of math anxiety. If you steer clear of math at all costs, then you can't set yourself up for the success you deserve.

Lack of Motivation

Students with math anxiety may experience a lack of motivation. They may struggle to find the incentive to get engaged with what they view as a frightening subject. These students are often overwhelmed, making it difficult for them to complete or even start math assignments.

Procrastination

Another symptom of math anxiety is procrastination. Students may voluntarily delay or postpone their classwork and assignments, even if they know there will be a negative consequence for doing so. Additionally, they may choose to wait until the last minute to start projects and homework, even when they know they need more time to put forth their best effort.

Physiological Reactions

Many people with a fear of math experience physiological side effects. These may include an increase in heart rate, sweatiness, shakiness, nausea, and irregular breathing. These symptoms make it difficult to focus on the math content, causing the student even more stress and fear.

Strong Emotional Responses

Math anxiety also affects people on an emotional level. Responding to math content with strong emotions such as panic, anger, or despair can be a sign of math anxiety.

Low Test Scores and Performance

Low achievement can be both a symptom and a cause of math anxiety. When someone does not take the steps needed to perform well on tests and assessments, they are less likely to pass. The more they perform poorly, the more they accept this poor performance as a fact that can't be changed.

Feeling Alone

People who experience math anxiety feel like they are the only ones struggling, even if the math they are working on is challenging to many people. Feeling isolated in what they perceive as failure can trigger tension or nervousness.

Feeling of Permanency

Math anxiety can feel very permanent. You may assume that you are naturally bad at math and always will be. Viewing math as a natural ability rather than a skill that can be learned causes people to believe that nothing will help them improve. They take their current math abilities as fact and assume that they can't be changed. As a result, they give up, stop trying to improve, and avoid engaging with math altogether.

Lack of Confidence

People with low self-confidence in math tend to feel awkward and incompetent when asked to solve a math problem. They don't feel comfortable taking chances or risks when problem-solving because they second-guess themselves and assume they are incorrect. They don't trust in their ability to learn the content and solve problems correctly.

Panic

A general sense of unexplained panic is also a sign of math anxiety. You may feel a sudden sense of fear that triggers physical reactions, even when there is no apparent reason for such a response.

Causes Of Math Anxiety

Math anxiety can start at a young age and may have one or more underlying causes. Common causes of math anxiety include the following:

The Attitude of Parents or Guardians

Parents often put pressure on their children to perform well in school. Although their intentions are usually good, this pressure can lead to anxiety, especially if the student is struggling with a subject or class.

Perhaps your parents or others in your life hold negative predispositions about math based on their own experiences. For instance, if your mother once claimed she was not good at math, then you might have incorrectly interpreted this as a predisposed trait that was passed down to you.

Teacher Influence

Students often pick up on their teachers' attitudes about the content being taught. If a teacher is happy and excited about math, students are more likely to mirror these emotions. However, if a teacher lacks enthusiasm or genuine interest, then students are more inclined to disengage.

Teachers have a responsibility to cultivate a welcoming classroom culture that is accepting of mistakes. When teachers blame students for not understanding a concept, they create a hostile classroom environment where mistakes are not tolerated. This tension increases student stress and anxiety, creating conditions that are not conducive to inquiry and learning. Instead, when teachers normalize mistakes as a natural part of the problem-solving process, they give their students the freedom to explore and grapple with the math content. In such an environment, students feel comfortable taking chances because they are not afraid of being wrong.

Students need teachers that can help when they're having problems understanding difficult concepts. In doing so, educators may need to change how they teach the content. Since different people have unique learning styles, it's the job of the teacher to adapt to the needs of each student. Additionally, teachers should encourage students to explore alternate problem-solving strategies, even if it's not the preferred method of the educator.

Fear of Being Wrong

Embarrassing situations can be traumatic, especially for young children and adolescents. These experiences can stay with people through their adult lives. Those with math anxiety may experience a fear of being wrong, especially in front of a group of peers. This fear can be paralyzing, interfering with the student's concentration and ability to focus on the problem at hand.

Timed Assessments

Timed assessments can help improve math fluency, but they often create unnecessary pressure for students to complete an unrealistic number of problems within a specified timeframe. Many studies have shown that timed assessments often result in increased levels of anxiety, reducing a student's overall competence and ability to problem-solve.

Debunking Math Myths

There are lots of myths about math that are related to the causes and development of math-related anxiety. Although these myths have been proven to be false, many people take them as fact. Let's go over a few of the most common myths about learning math.

Myth: Men Are Better at Math Than Women

Math has a reputation for being a male-dominant subject, but this doesn't mean that men are inherently better at math than women. Many famous mathematical discoveries have been made by women. Katherine Johnson, Dame Mary Lucy Cartwright, and Marjorie Lee Brown are just a few of the many famous women mathematicians. Expecting to be good or bad at math because of your gender sets you up for stress and confusion. Math is a skill that can be learned, just like cooking or riding a bike.

Myth: There Is Only One Good Way to Solve Math Problems

There are many ways to get the correct answer when it comes to math. No two people have the same brain, so everyone takes a slightly different approach to problem-solving. Moreover, there isn't one way of problem-solving that's superior to another. Your way of working through a problem might differ from someone else's, and that is okay. Math can be a highly individualized process, so the best method for you should be the one that makes you feel the most comfortable and makes the most sense to you.

Myth: Math Requires a Good Memory

For many years, mathematics was taught through memorization. However, learning in such a way hinders the development of critical thinking and conceptual understanding. These skill sets are much more valuable than basic memorization. For instance, you might be great at memorizing mathematical formulas, but if you don't understand what they mean, then you can't apply them to different scenarios in the real world. When a student is working from memory, they are limited in the strategies available to them to problem-solve. In other words, they assume there is only one correct way to do the math, which is the method they memorized. Having a variety of problem-solving options can help students figure out which method works best for them. Additionally, it provides students with a better understanding of how and why certain mathematical strategies work. While memorization can be helpful in some instances, it is not an absolute requirement for mathematicians.

Myth: Math Is Not Creative

Math requires imagination and intuition. Contrary to popular belief, it is a highly creative field. Mathematical creativity can help in developing new ways to think about and solve problems. Many people incorrectly assume that all things are either creative or analytical. However, this black-and-white view is limiting because the field of mathematics involves both creativity and logic.

Myth: Math Isn't Supposed To Be Fun

Whoever told you that math isn't supposed to be fun is a liar. There are tons of math-based activities and games that foster friendly competition and engagement. Math is often best learned through play, and lots of mobile apps and computer games exemplify this.

Additionally, math can be an exceptionally collaborative and social experience. Studying or working through problems with a friend often makes the process a lot more fun. The excitement and satisfaction of solving a difficult problem with others is quite rewarding. Math can be fun if you look for ways to make it more collaborative and enjoyable.

Myth: Not Everyone Is Capable of Learning Math

There's no such thing as a "math person." Although many people think that you're either good at math or you're not, this is simply not true. Everyone is capable of learning and applying mathematics. However, not everyone learns the same way. Since each person has a different learning style, the trick is to find the strategies and learning tools that work best for you. Some people learn best through hands-on experiences, and others find success through the use of visual aids. Others are auditory learners and learn best by hearing and listening. When people are overwhelmed or feel that math is too hard, it's often because they haven't found the learning strategy that works best for them.

Myth: Good Mathematicians Work Quickly and Never Make Mistakes

There is no prize for finishing first in math. It's not a race, and speed isn't a measure of your ability. Good mathematicians take their time to ensure their work is accurate. As you gain more experience and practice, you will naturally become faster and more confident.

Additionally, everyone makes mistakes, including good mathematicians. Mistakes are a normal part of the problem-solving process, and they're not a bad thing. The important thing is that we take the time to learn from our mistakes, understand where our misconceptions are, and move forward.

Myth: You Don't Need Math in the Real World

Our day-to-day lives are so infused with mathematical concepts that we often don't even realize when we're using math in the real world. In fact, most people tend to underestimate how much we do math in our everyday lives. It's involved in an enormous variety of daily activities such as shopping, baking, finances, and gardening, as well as in many careers, including architecture, nursing, design, and sales.

Tips and Strategies for Overcoming Math Anxiety

If your anxiety is getting in the way of your level of mathematical engagement, then there are lots of steps you can take. Check out the strategies below to start building confidence in math today.

Focus on Understanding, Not Memorization

Don't drive yourself crazy trying to memorize every single formula or mathematical process. Instead, shift your attention to understanding concepts. Those who prioritize memorization over conceptual understanding tend to have lower achievement levels in math. Students who memorize may be able to complete some math, but they don't understand the process well enough to apply it to different situations. Memorization comes with time and practice, but it won't help alleviate math anxiety. On the other hand, conceptual understanding will give you the building blocks of knowledge you need to build up your confidence.

Replace Negative Self-Talk with Positive Self-Talk

Start to notice how you think about yourself. Whenever you catch yourself thinking something negative, try replacing that thought with a positive affirmation. Instead of continuing the negative thought, pause to reframe the situation. For ideas on how to get started, take a look at the table below:

Instead of thinking...	Try thinking...
"I can't do this math." "I'm not a math person."	"I'm up for the challenge, and I'm training my brain in math."
"This problem is too hard."	"This problem is hard, so this might take some time and effort. I know I can do this."
"I give up."	"What strategies can help me solve this problem?"
"I made a mistake, so I'm not good at this."	"Everyone makes mistakes. Mistakes help me to grow and understand."
"I'll never be smart enough."	"I can figure this out, and I am smart enough."

Practice Mindfulness

Practicing mindfulness and focusing on your breathing can help alleviate some of the physical symptoms of math anxiety. By taking deep breaths, you can remind your nervous system that you are not in immediate danger. Doing so will reduce your heart rate and help with any irregular breathing or shakiness. Taking the edge off of the physiological effects of anxiety will clear your mind, allowing your brain to focus its energy on problem-solving.

Do Some Math Every Day

Think about learning math as if you were learning a foreign language. If you don't use it, you lose it. If you don't practice your math skills regularly, you'll have a harder time achieving comprehension and fluency. Set some amount of time aside each day, even if it's just for a few minutes, to practice. It might take some discipline to build a habit around this, but doing so will help increase your mathematical self-assurance.

Use All of Your Resources

Everyone has a different learning style, and there are plenty of resources out there to support all learners. When you get stuck on a math problem, think about the tools you have access to, and use them when applicable. Such resources may include flashcards, graphic organizers, study guides, interactive notebooks, and peer study groups. All of these are great tools to accommodate your individual learning style. Finding the tools and resources that work for your learning style will give you the confidence you need to succeed.

Realize That You Aren't Alone

Remind yourself that lots of other people struggle with math anxiety, including teachers, nurses, and even successful mathematicians. You aren't the only one who panics when faced with a new or challenging problem. It's probably much more common than you think. Realizing that you aren't alone in your experience can help put some distance between yourself and the emotions you feel about math. It also helps to normalize the anxiety and shift your perspective.

Ask Questions

If there's a concept you don't understand and you've tried everything you can, then it's okay to ask for help! You can always ask your teacher or professor for help. If you're not learning math in a traditional classroom, you may want to join a study group, work with a tutor, or talk to your friends. More often than not, you aren't the only one of your peers who needs clarity on a mathematical concept. Seeking understanding is a great way to increase self-confidence in math.

Remember That There's More Than One Way To Solve A Problem

Since everyone learns differently, it's best to focus on understanding a math problem with an approach that makes sense to you. If the way it's being taught is confusing to you, don't give up. Instead, work to understand the problem using a different technique. There's almost always more than one problem-solving method when it comes to math. Don't get stressed if one of them doesn't make sense to you. Instead, shift your focus to what does make sense. Chances are high that you know more than you think you do.

Visualization

Visualization is the process of creating images in your mind's eye. Picture yourself as a successful, confident mathematician. Think about how you would feel and how you would behave. What would your work area look like? How would you organize your belongings? The more you focus on something, the more likely you are to achieve it. Visualizing teaches your brain that you can achieve whatever it is that you want. Thinking about success in mathematics will lead to acting like a successful mathematician. This, in turn, leads to actual success.

Focus on the Easiest Problems First

To increase your confidence when working on a math test or assignment, try solving the easiest problems first. Doing so will remind you that you are successful in math and that you do have what it takes. This process will increase your belief in yourself, giving you the confidence you need to tackle more complex problems.

Find a Support Group

A study buddy, tutor, or peer group can go a long way in decreasing math-related anxiety. Such support systems offer lots of benefits, including a safe place to ask questions, additional practice with mathematical concepts, and an understanding of other problem-solving explanations that may work better for you. Equipping yourself with a support group is one of the fastest ways to eliminate math anxiety.

Reward Yourself for Working Hard

Recognize the amount of effort you're putting in to overcome your math anxiety. It's not an easy task, so you deserve acknowledgement. Surround yourself with people who will provide you with the positive reinforcement you deserve.

Remember, You Can Do This!

Conquering a fear of math can be challenging, but there are lots of strategies that can help you out. Your own beliefs about your mathematical capabilities can limit your potential. Working toward a growth mindset can have a tremendous impact on decreasing math-related anxiety and building confidence. By knowing the symptoms of math anxiety and recognizing common misconceptions about learning math, you can develop a plan to address your fear of math. Utilizing the strategies discussed can help you overcome this anxiety and build the confidence you need to succeed.

Thank You

We at Mometrix would like to extend our heartfelt thanks to you, our friend and patron, for allowing us to play a part in your journey. It is a privilege to serve people from all walks of life who are unified in their commitment to building the best future they can for themselves.

The preparation you devote to these important testing milestones may be the most valuable educational opportunity you have for making a real difference in your life. We encourage you to put your heart into it—that feeling of succeeding, overcoming, and yes, conquering will be well worth the hours you've invested.

We want to hear your story, your struggles and your successes, and if you see any opportunities for us to improve our materials so we can help others even more effectively in the future, please share that with us as well. **The team at Mometrix would be absolutely thrilled to hear from you!** So please, send us an email (support@mometrix.com) and let's stay in touch.

If you'd like some additional help, check out these other resources we offer for your exam:
http://MometrixFlashcards.com/CCSS

Online Resources

Due to our efforts to try to keep this book to a manageable length, we've created a link that will give you access to all of your online resources:

mometrix.com/resources719/cchsstatprob

It's Your Moment, Let's Celebrate It!

Share your story @mometrixtestpreparation